An Introduction to Systematic Geomorphology

VOLUME FOUR

COASTS

An Introduction to Systematic Geomorphology

VOLUME FOUR

COASTS

E. C. F. Bird

THE M.I.T. PRESS

Massachusetts Institute of Technology

Cambridge, Massachusetts, and London, England

First published 1968 by
The Australian National University Press
Canberra, Australia

Second M.I.T. Press Printing, November 1970

ISBN 0 262 02050 5 (hardcover)

Library of Congress Catalog Card No. 68-27425

INTRODUCTION TO THE SERIES

This series is conceived as a systematic geomorphology at university level. It will have a role also in high school education and it is hoped the books will appeal as well to many in the community at large who find an interest in the why and wherefore of the natural scenery around them.

The point of view adopted by the authors is that the central themes of geomorphology are the characterisation, origin, and evolution of landforms. The study of processes that make landscapes is properly a part of geomorphology, but within the present framework process will be dealt with only in so far as it elucidates the nature and history of the landforms under discussion. Certain other fields such as submarine geomorphology and a survey of general principles and methods are also not covered in the volumes as yet planned. Some knowledge of the elements of geology is presumed.

Four volumes will approach landforms as parts of systems in which the interacting processes are almost completely motored by solar energy. In humid climates (Volume One) rivers dominate the systems. Fluvial action, operating differently in some ways, is largely responsible for the landscapes of deserts and savanas also (Volume Two), though winds can become preponderant in some deserts. In cold climates, snow, glacier ice, and ground ice come to the fore in morphogenesis (Volume Three). On coasts (Volume Four), waves, currents, and wind are the prime agents in the complex of processes fashioning the edge of the land.

Three further volumes will consider the parts played passively by the attributes of the earth's crust and actively by processes deriving energy from its interior. Under structural landforms (Volume Five), features immediately consequent on earth movements and those resulting from tectonic and lithologic guidance of denudation are considered. Landforms directly the product of volcanic activity and those created by erosion working on volcanic materials are sufficiently distinctive to warrant separate treatment (Volume Six). Though karst is undoubtedly delimited lithologically, it is fashioned by a special combination of processes centred on solution so that the seventh volume partakes also of the character of the first group of volumes.

J. N. JENNINGS
General Editor

PREFACE

This book is designed as an introduction to the subject of coastal geomorphology for high school and university students and interested laymen. It is based on courses given by the author in recent years in the Australian National University and the Universities of London, Sydney, and Melbourne, and follows a similar arrangement to a previous textbook written for Australian students and entitled *Coastal Landforms: An introduction to coastal geomorphology with Australian examples* (Australian National University Press, 1964), the aim being to provide an introductory account of the data and problems of coastal geomorphology with examples from various parts of the world. Inevitably there is some bias towards coasts that the author knows at first hand in the British Isles, parts of western Europe, the United States, and Australia, but an effort has been made to draw upon published work on other areas to balance this.

Definitions of terms, given when these are first encountered in the text, can be located by using the index. As a rule the terms and definitions used are those commonly accepted by coastal geomorphologists, but more advanced students interested in the origins and historical connotations of coastal terminology will need to consult the standard geological and geographical glossaries. The selected bibliography is intended to lead advanced students into the substantial research literature of the subject.

The metric system is used in preference to other units of measurement in order to bring the subject in line with customary scientific practice. Unfortunately, much of the basic data in the earth sciences is still recorded and reported in terms of feet, yards, miles and other non-metric units, even in some of the scientific journals, and in coastal work there is the additional complication that charts give soundings in fathoms, distances in nautical miles, and current velocities in knots. Conversion to metric units raised problems when the measurements quoted were approximations, and an effort has been made to render such measurements in metric units without implying different orders of accuracy from those originally intended. Where confusion is possible, non-metric alternatives have been included.

I would like to thank Mr J. N. Jennings, editor of the Introduction to Systematic Geomorphology series, for critical discussion of the manuscript and for providing the photograph of the Nullarbor coastline which is used on the jacket; Mr Peter Daniell, who drew the maps; and the various people and organisations who contributed the plates, as acknowledged in the respective captions.

E. C. F. B.

Melbourne,
October 1967.

CONTENTS

FIGURES

PLATES

I

INTRODUCTION

Coastal geomorphology is the study of coastal landforms, their evolution, the processes at work on them, and the changes now taking place.

It is convenient to begin by defining the terms used to describe coastal features. The shore (Figs. 1 and 2) is the zone between the water's edge at normal low tide and the landward limit of effective

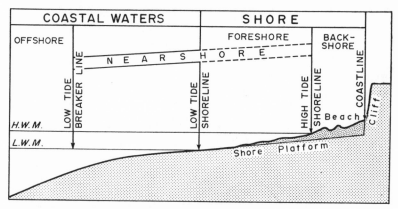

1   *Geomorphological terminology for a cliffed coast*

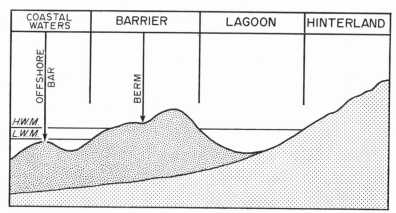

2   *Geomorphological terminology for a depositional coast*

1

wave action; it comprises the foreshore, exposed at low tide and submerged at high tide, and the backshore, extending above normal high tide level, but inundated by exceptionally high tides or by large waves during storms. The shoreline is strictly the water's edge, migrating to and fro with the tide, and the nearshore zone lies between this and the line where waves begin to break. Outside the breaker line, extending to an arbitrary limit in deep water, is the offshore zone, but the term offshore is also used, together with onshore and longshore, to describe directions of flow of wind, water, and sediment.

A beach is an accumulation of loose sediment, such as sand, shingle, and boulders, sometimes confined to the backshore but often extending across the foreshore as well. A bank of beach material which lies offshore and is exposed at high tide is called a barrier, but a similar feature submerged for at least part of the tidal cycle is called a bar. A spit is a beach that diverges from the coast, often terminating in one or more landward projections, known as recurves.

The coast is a zone of varying width, including the shore and extending to the landward limit of penetration of marine influences: the crest of a cliff, the head of a tidal estuary, or the solid ground that lies behind coastal dunes, lagoons, and swamps. The coastline is usually taken as the land margin in the backshore zone. It is useful to refer to the sea area adjoining the coast, comprising the nearshore and offshore zones, as coastal waters.

Coastal evolution

The geomorphological processes at work on coastal landforms are influenced by a number of environmental factors, notably geological, climatic, biotic, and tidal and other oceanographic factors, including salinity. These vary from one sector of the coast to another, the variation being zonal in terms of climatic regions, and irregular in terms of geological outcrops.

The geological factor is obvious in the evolution of cliffed coasts, which show features related to the structure and lithology of rock formations which make up the hinterland, the coast, and the nearshore zone. Depositional coasts are also influenced by the geological factor in terms of the sources of sediment derived from adjacent cliffed coasts, hinterland drainage basins, or the sea floor.

The climatic factor is important in terms of the weathering of coastal rock outcrops, which results from physical, chemical, and biological processes related partly to subaerial climatic conditions and partly to the presence or proximity of the sea. Rocks are decomposed or disintegrated by such processes as repeated wetting and drying, solution by rain water, thermal expansion and contraction and freeze-thaw alternations, all related to temperature, precipitation, and evaporation régimes in the coastal environment. Climate also conditions modes of subaerial erosion, coastal slopes washed by runoff showing various kinds of wastage ranging from soil creep and solifluction to gulleying, landslides, and mudflows, but marine erosion plays an important part in undercutting coastal slopes, initiating instability, and removing superficial materials to expose fresh rock outcrops for further subaerial weathering. Regional variations in climate (Bailey, 1958)* are marked by contrasted modes of geomorphological evolution on coasts. In the humid tropics, rapid chemical weathering at high temperatures results in deep decomposition of many of the rock formations which outcrop at the coast, and bold headlands are restricted to the formations which resist such weathering, notably quartzites. Rivers draining these areas generally deliver an abundance of fine sediment (silt and clay) to the coast, rather than the sands and gravels which the same rock outcrops would yield under the weathering régimes of cooler or drier environments. In cold regions, by contrast, cliffed coasts are modified by rock shattering and solifluction resulting from repeated alternations of freezing and thawing, which yield an abundance of rock fragments for the building of beaches, spits, and barriers of shingle along the coast. Where glaciers reach the coast, glacifluvial and morainic material is delivered to the shore, while in high latitudes ice coasts form a distinct category. In arid regions the limited runoff from the land carries only small quantities of terrigenous sediment to the coast, so that many of the beach and barrier systems consist largely of biogenic sediment derived from marine shell and coral debris.

The biotic factor is strongly influenced by climatic conditions which limit the range of many significant organisms. Corals and associated reef-building organisms are confined mainly to the inter-tropical zone, and mangrove swamps are restricted to estuaries and sheltered coastal environments in low latitudes—environ-

References are listed on pp. 227–37.

ments which are occupied by salt marshes in the temperate zone. The effects of organisms may be erosional, protective, or constructional. The shore fauna and flora often include species, the growth and metabolism of which lead to the decomposition or dissolution of coastal rock outcrops, especially on limestones. On the other hand, salt marsh and dune vegetation may stabilise depositional landforms on the coast, or promote accretion of sediment in such a way as to build new landforms, and a variety of reef formations can be built by coral, algae, oysters, and other calcareous organisms on the shore and in coastal waters.

Wind is a climatic factor of particular importance in coastal evolution, building coastal dunes and generating the waves and currents that, together with tides, create the pattern of nearshore water circulation which influences the sculpturing of cliffed coasts, shaping of sea-floor topography, movement of coastal sediment, and patterns of marine deposition. Variations in the strength and frequency of winds result in contrasts in modes of coastal evolution. Some coasts border stormy seas; others face comparatively calm waters. Some receive ocean swell, generated by distant storms and transmitted across the oceans; others are protected from prevailing ocean conditions by promontories and islands, depositional barriers, or offshore reefs and shoals, and receive only the waves generated locally by winds blowing over coastal waters.

Tidal conditions also vary from coast to coast. On some coasts the tide range is negligible; on others it is measured in tens of feet, and strong ebb and flow currents are generated in the coastal environment.

The oceanographic factor derives from the nature of sea water, with variations in salinity from almost fresh, in parts of the Baltic Sea, to relatively strong salinity in the Red Sea and other ocean areas in the arid zone. Salt water and sea spray have corrosive effects which influence the weathering of coastal rock outcrops; they also produce distinctive ecological conditions, the habitats of marine and estuarine flora and fauna, many of which influence weathering, erosion, transportation, and deposition of rock materials in the coastal environment.

Coastal evolution can be considered in terms of morphogenic systems, within which these various factors influence the geomorphological processes acting upon the coast. There is 'feedback' in the sense that the developing morphology influences the geomorphological processes, and becomes one of the factors influencing subsequent coastal evolution. Variations in morphogenic systems on the world's coasts have been outlined by Davies (1964), who advocated the use of this concept in coastal studies. In practice it is necessary to bear in mind two further factors in coastal landform studies: the possible existence of relict features, inherited from contrasted morphogenic conditions in the past, and modifications of coastal systems by human activities during the past few centuries.

Features related to past conditions are found on many coasts. They include landforms which developed when the sea stood higher, relative to the land, and features related to submergence following relatively low sea level phases. There are coastal landforms inherited from earlier episodes when the climate was warmer or cooler, wetter or drier, than it is now. The existing morphogenic systems were established only within the last few thousand years, during Recent (Holocene) times, and the landform legacy of the preceding Pleistocene period, when marked variations of climate and sea level accompanied the waxing and waning of ice sheets over a period of at least a million years, is still evident on many coasts.

Modification of coastal landforms by human activities may be either direct or indirect. The building of sea walls, groynes, and breakwaters, the dredging of harbour entrances, and the dumping of material on the coast and offshore, are all direct modifications of coastal topography, but human interference often has indirect effects, accentuating erosion or deposition on adjacent sectors of the coast. Reduction of coastal vegetation by cutting, burning, grazing, or pollution has often led to changes in patterns of erosion and deposition on the coast, in estuaries and lagoons, or on the sea floor.

Attempts to modify coastal changes—to halt erosion or prevent the silting of a harbour entrance—require an understanding of the factors and processes at work in the coastal morphogenic system: the pattern of change, the sources of sediment, the paths of sediment flow, and the quantities involved over given periods of time. Such knowledge enables the effects of various kinds of human interference to be predicted and offset, or allowed for, in coastal engineering works. Studies of the effects of human activities on the coast are also important in coastal land management and in the conservation of natural resources, including scenery and wildlife, in coastal areas.

The need for accurate data has introduced new techniques to coastal geomorphology, supplementing traditional methods of observation, mapping, and measurement of the landforms, processes, and changes on the coast. Aerial photography has been used for some time as an aid to mapping and measurement of coastal changes, but modern developments in colour photography from the air are extending these studies to the sea floor, and satellite photography holds possibilities for synoptic recording of ocean wave conditions. Aqualung diving as been used to supplement the older methods of sounding and sampling in the offshore zone. Depositional formations on the coast are subjected to modern stratigraphical and sedimentological analyses, using radiometric as well as palaeontological and archaeological methods of dating, and tracers are used to follow and measure currents and coastal sediment flow. Scale models of coastal areas have their limitations in practice, but have proved useful in investigating certain coastal phenomena.

Each of these topics will be considered in more detail in subsequent chapters. Before proceeding to describe the various kinds of coastal landforms it is necessary to examine the processes at work in coastal waters (Chapter II) and the evidence of past changes of land and sea levels, which preceded the development of the environments in which we now find coastal landforms (Chapter III).

II
TIDES, WAVES, AND CURRENTS

Tides

Tides are movements of the oceans set up by the gravitational effects of the sun and the moon in relation to the earth. They are important in coastal geomorphology because they lead to regular changes in the level of the sea along the coast, and because currents are generated as the tide ebbs and flows. It is not necessary here to consider in detail the nature and origin of tides; they are essentially oscillations of water in ocean basins with characteristics determined partly by the size and shape of the basin. They move in harmony with the gravitational forces of the sun and the moon, with an additional gyratory motion imparted by the earth's rotation (Russell and MacMillan, 1954). They circulate around a number of nodal (amphidromic) points, and their effects are most strongly marked in shallow and relatively enclosed ocean areas such as the English Channel and the North Sea bordering the British Isles.

The rise and fall of the tide on a coast is measured by tide gauges, located chiefly at ports. The tides recorded shortly after new moon and full moon, when earth, sun, and moon are in alignment, and have combined gravitational effects, are relatively large, and are known as spring tides. Maximum spring tide ranges occur about the equinoxes (late March and late October), when the sun is overhead at the equator. At half-moon (first and last quarter), when the sun and moon are at right angles in relation to the earth, their gravitational effects are not combined; tide ranges recorded shortly after this are reduced, and are known as neap tides. There are also long-term variations in tide range in response to astronomical cycles of the relative positions of sun, moon, and earth; the moon, for example, returns to a similar position relative to the earth once every 27·5 days, but its orbit is such that it returns to exactly the same position only once in 18·6 years. Accurate determination of mean sea level from tide gauge records therefore requires tidal observations over a longer period than this, but a reasonable estimate can be made on the basis of one or two years' observations.

Further complications are introduced by meteorological effects. Onshore winds build up tide levels in coastal waters whereas offshore winds lower them. A fall in atmospheric pressure is accompanied by elevation, and a rise by depression, of the ocean surface: a fall of 1 millibar raises the ocean level about 1 cm. Storm surges are produced when strong onshore winds build up coastal water to an exceptionally high level, and are most serious when the onshore winds coincide with high spring tides. In 1953 a storm surge produced by a northerly gale in the North Sea raised the level of coastal water about 3 m, flooding extensive areas in E England and low-lying parts of the Dutch and German coasts (Steers, 1953b). High tides augmented by meteorological effects are known as king tides on the N coast of Australia, where heavy rains accompany onshore winds during the summer monsoon and cause widespread flooding in coastal districts. Hurricanes raise sea level as much as 6 m along the shores of the Gulf of Mexico, and disastrous storm surges are experienced from time to time at the head of the Bay of Bengal, when large areas of the Sundarbans are devastated by submergence.

Observations of ocean levels during the International Geophysical Year (1957–8) showed fluctuations distinct from changes that could be explained by weather conditions, and these fluctuations are thought to be the result of seasonal variations in the temperature and salinity of the oceans. The volume of a mass of sea water increases when rising temperature or diminishing salinity change its density, and this increase results in a slight rise in sea level. There are also local variations in tide range due to the configuration of the coast, the effects of outflow from rivers, and the modifications introduced where breakwaters and jetties have been built at harbour entrances. It is therefore necessary to consider the situation of a tide gauge when deciding the extent of coast to which a particular tide record is applicable.

Tide range in mid-ocean is small, of the order of a $\frac{1}{2}$ m, but it increases where the tide invades shallow coastal waters, particularly in gulfs and embayments. Where the spring tide range is less than 2 m, the environment is termed *microtidal;* between 2 and 4 m *mesotidal;* and more than 4 m *macrotidal.* Microtidal ranges are typical of the coasts of the Atlantic, Pacific, Indian, and Southern Oceans and of certain landlocked seas (Baltic, Mediterranean, Black, Red, and Caribbean Seas). Around the Pacific Basin mean spring tide range exceeds 2 m (about 6 feet) only in certain gulfs, notably the Gulf of Siam (Bangkok, 2·4 m (7·8 feet)) and the Gulf of California (9 m (about 30 feet) at the mouth of the Colorado River). On the Australian coast microtidal conditions prevail from Brisbane S to Cape Howe, thence along the whole of the S and W coasts, with the exception of certain embayments. Tide range increases into Spencer Gulf (Port Augusta, 2·0 m (6·6 feet)) and Gulf St Vincent (Port Wakefield, 2·1 m (7 feet)), but the narrow entrance to Port Phillip Bay (Fig. 8, p. 23) impedes tidal invasion, and tide ranges within the bay (0·85 m = 2·8 feet at Port Melbourne) are smaller than those outside (1·47 m = 4·8 feet at the entrance). Macrotidal ranges are characteristic of the N coast of Australia, especially between Port Hedland (up to 5·8 m = 19 feet) and Darwin (up to 5·5 m = 18 feet), with very large ranges recorded in the gulfs; Derby, on the shore of King Sound, has a spring tide range of 10 m (33 feet). Macrotidal areas are also found in the Bay of St Michel in NW France, where maximum tides exceed 9 m (30 feet), the Bristol Channel, where mean spring tide range increases eastwards to 12·3 m (40·3 feet) at Avonmouth, and the head of the Bay of Fundy, where the world's largest spring tide range, more than 15·2 m (50 feet), is recorded.* Where large tides enter estuaries a visible wave, known as a tidal bore, may develop as the tidal front steepens (Tricker, 1964). In the Amazon River the tidal bore, migrating upstream, attains a height of about 9 m. Little is known of the dimensions of tidal bores developed in rivers draining to the marine gulfs of the N coast of Australia, but the bore that develops in the Victoria River is said to run at least fifty miles upstream from Joseph Bonaparte Gulf.

Contrasts in tide range have important consequences in coastal geomorphology (Davies, 1964). A large tide range implies a broad inter-tidal zone, more than 20 km of sandflats and mudflats being exposed at low spring tides in the Bay of St Michel. Wave energy is expended in traversing such a broad shore zone and the waves which reach the backshore at high tide are much diminished by friction in their shallow water passage. Cliff erosion and beach building proceed only intermittently during the brief high tide, wave action being withdrawn from the backshore for most of the

* Tidal information (given in feet) may be obtained from the three volumes of *Admiralty Tide Tables* (I. *European Waters and Mediterranean*, II. *Atlantic and Indian Oceans*, III. *Pacific Ocean and Adjacent Seas*) published annually in London by the Hydrographer of the Navy.

tidal cycle. On sheltered coasts with a macrotidal range, as in Bridgwater Bay in the Bristol Channel, extensive salt marshes may develop. Where tide range is small, wave energy is concentrated at a more consistent level, facilitating cliff erosion and impeding the development of coastal marshlands. Larger tide ranges tend also to generate stronger tidal currents (see below, p. 22).

Waves

Waves are superficial undulations of the water surface produced by winds blowing over the sea (Bascom, 1959). The turbulent flow of wind energy over water sets up stress and pressure variations on the sea surface initiating waves which grow as the result of a pressure contrast that develops between their windward and leeward slopes. Waves consist of orbital movements of water which diminish rapidly from the surface downwards, until motion is very small ($0.04 \times$ surface orbital diameter) where the water depth (d) equals half the wave length (L) (Fig. 3). In fact the orbital motion is not quite complete, so that water particles move forward slightly as each wave passes, producing a slight drift of water in the direction of wave advance. Waves are essentially a means of transmitting energy through water with relatively small displacement of water particles in the direction of energy flow.

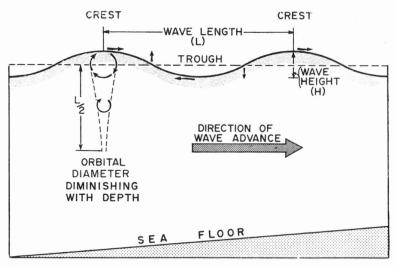

3   *Wave terminology*

In deep water, wave velocity (C_O), the rate of movement of a wave crest in metres per second, is the ratio of wave length (L_O, measured in metres) to wave period, the time interval between the passage of successive wave crests (T, measured in seconds). Wave velocity can be calculated from the following formula, in which g represents gravitational acceleration (approximately 980·62 cm/sec² at latitude 45°):

$$C_O{}^2 = \frac{gL_O}{2\pi}$$

Or, since $L_O = C_O T$,

$$C_O = \frac{gT}{2\pi} = 1\cdot56 \text{ T in metres per second.}$$

From this it is evident that

$$L_O = 1\cdot56T^2$$

and it is convenient to calculate wave length from measurement of wave period by using this formula.

Wave height (H) is the vertical distance between adjacent crests and troughs, and wave steepness is conventionally the ratio between the height and the length: H_O/L_O. The dimensions of waves are determined partly by wind velocity, partly by fetch (the extent of open water across which the wind is blowing), and partly by the duration of the wind. In coastal waters where fetch is limited the height of the waves is proportional to wind velocity and the wave period to the square root of wind velocity. Observations from ships in mid-ocean suggest that the largest waves, generated by prolonged strong winds over fetches of at least 500 km, may be more than 20 m high, and may travel at more than 80 km/hr. There is usually considerable variation in the dimensions of waves reaching the shore, but the difficulty of estimating 'typical' wave lengths and wave heights can be overcome by using the concept of the significant wave. This is based on the calculation of the mean wave length ($L_\frac{1}{3}$) and the mean wave height ($H_\frac{1}{3}$) of the highest one-third of all waves observed over a period of 20 minutes. The dimensions thus obtained can be used in comparisons of conditions in coastal waters at different times, or on different coastal sectors.

As waves move into shallow coastal waters ($d < \frac{L}{2}$) they are modified in several ways. Their velocity (C_S in shallow water) is diminished according to the formula

$$C_S{}^2 = \frac{gL}{2\pi} \cdot \tanh \frac{2\pi d}{L}$$

As velocity diminishes, so do wave length (L_S) and period (T). Wave height first diminishes as the ratio of water depth to wave length ($\frac{d}{L}$) falls below 0·5, but when this ratio becomes less than 0·06 there is a revival of wave height, the crest rising immediately before breaking. Correspondingly, wave steepness falls at first, then rises, the wave crest becoming narrower and sharper, the trough wider and flatter. Orbital movements within the wave become more and more elliptical, and shoreward velocity in the wave crest increases until it is greater than the wave velocity. The orbital motion can then no longer be completed, and the unsupported wave front collapses sending forth a rush of water, known as the swash, on to the shore. Breakers are said to plunge when the wave front curves over and collapses with a crash, or to spill, when the crest breaks more gradually into a swash running forward on to the shore; the way in which this happens depends partly on the gradient and topography of the nearshore sea floor. The swash is followed by a withdrawal, the backwash, and water that has been carried shoreward by wave action returns to the sea either as undertow (sheet flow near the sea bed) or in localised rip currents (Shepard *et al.*, 1941) with water flowing back through

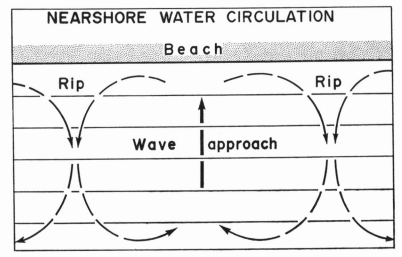

4   *Nearshore water circulation with waves approaching parallel to the shoreline*

the breaker line in sectors up to 30 m wide, and attaining velocities of up to 8 km/hr before dispersing seaward. These various movements associated with wave action are known as the nearshore water circulation (Fig. 4).

Rip currents occur in definite patterns on many beaches. A light or moderate swell produces numerous rip current systems, and a heavy swell produces a few concentrated rips, fed by strong lateral currents in the surf zone (McKenzie, 1958). When waves arrive at an angle to the shoreline they deflect the nearshore water circulation, so that resultant currents travel in one direction along the shore. The effects of these wave-induced longshore currents are not easily distinguished from the direct effects of waves arriving at an angle to the shore and breaking on the beach, for both processes move sediment along the shore, the action of oblique wave swash causing lateral *beach drifting*, while the wave-induced currents cause *longshore drifting* in the nearshore zone (see Chapter V). Rip currents that develop when waves approach obliquely head away diagonally through the surf instead of straight out to sea (Fig. 5).

A distinction is made between irregular patterns of waves related to locally existing wind conditions (a 'sea' in nautical terminology), and waves of more distant origin arriving at the coast as an ocean swell. Storms in mid-ocean produce a spectrum of waves of various lengths and heights, fanning out 30°–40° on either side of the generating wind direction and travelling in a Great Circle course across the oceans (Davies, 1964). As long waves travel faster than short waves and low waves persist longer than high waves (which have to overcome greater internal friction), the long, low waves travel far beyond the storm region as a relatively regular ocean swell. This is the swell that arrives on the Pacific coast of New South Wales on a calm day with wave periods of 12–16 seconds; the wave crests gain in height and steepness as they enter shallow water, and break to produce the surf that is famous on the Sydney beaches.

Similar swells are observed on the shores of the Pacific, Atlantic, and Indian Oceans, and may have travelled vast distances. It has been shown that an ocean swell arriving on the shores of San Clemente Island on the Californian coast originated in storm regions of the Southern Ocean, SW of Australia and New Zealand; it had travelled more than 16,000 km from the source region across

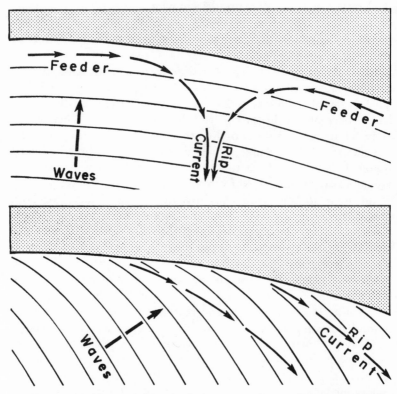

5 *Rip current patterns developed when waves arrive parallel to the shoreline
    (top) and at an angle to the shoreline (bottom) (after McKenzie, 1958)*

the Pacific Ocean to the coast of California (Munk *et al.*, 1963).
The swells that arrive on the S coast of Australia, passing into
Bass Strait and reaching the W and S coasts of Tasmania, also
originate in storms in the 'Roaring Forties', as do those reaching
the W coasts of Africa and South America. Storms in the Atlantic
Ocean produce waves typically of 6–8 seconds period which reach
the coasts of W Europe, and occasionally the shores of SW
England are washed by waves that have travelled more than
11,500 km from storm centres in the S half of the Atlantic Ocean.

Waves generated by wind action in coastal waters are typically
shorter (wave period less than 6 seconds) and less regular than
ocean swell; they may be superimposed on ocean swell in coastal
waters, an onshore wind accentuating the swell and adding shorter

waves to it and a cross wind producing shorter waves which move at an angle through the pattern of swell. Offshore winds tend to 'damp down' ocean swell, producing relatively calm conditions in the nearshore zone. Sectors of ocean coast protected by promontories, reefs, or offshore islands receive swells in a much modified and weakened form, and may even be exempt from them. In these conditions, as in the landlocked seas of the Mediterranean and the Baltic, locally wind-generated waves predominate, influencing shore processes and coastal evolution. Around the British Isles, wave régimes are largely determined by wind conditions in coastal waters, ocean swell penetrating the English Channel, the Irish Sea, and the North Sea only on a very limited scale. On Vancouver Island, off W Canada, there are marked contrasts between the Pacific coast, which receives ocean swell as well as waves generated by local winds, and the shores of the Strait of Georgia, where wave action is related entirely to local wind conditions. A large portion of the Australian coast is subject to the effects of ocean swell, the exceptions being the coast of Queensland where barrier reef structures lie offshore, and the coast of the Gulf of Carpentaria, partly sheltered from the ocean by Arnhem Land and the Cape York peninsula. Because of the narrowness of the entrance, ocean swell does not enter Port Phillip Bay (Fig. 7).

Ocean swell has parallel wave crests in deep water, but as the waves move into water shallower than the wave length they begin to 'feel bottom' and are modified in the manner previously described. At first the pattern of wave crests is only slightly modified, but where the depth diminishes to less than half the wave length and the free orbital motion of water is impeded, the frictional effects of the sea floor retard the advancing waves. Sea floor topography thus influences the pattern of swell approaching the coast, bending the wave crests until they are parallel to the submarine contours. This is known as wave refraction. Where the angle between the swell and the submarine contours is initially large, this adjustment is often incomplete by the time the waves arrive at the shore, so that they break at an angle (usually less than 10°); where the angular difference is small, the waves are refracted in such a way that they anticipate and fit the outline of the shore, breaking simultaneously along its length. Waves entering a broad embayment become refracted into gently curved patterns, the wave in the middle of the bay moving on in deeper water while, towards the sides, in shallower water, it is held back (Fig. 6A). Sharp irregularities of the sea floor have stronger effects: a submerged bank holds back the advancing waves, but a submarine trough allows them to run on (Fig. 6B). Emerged features, such as islands, or reefs awash at low tide, produce complex patterns of wave refraction, and waves that have passed through narrow straits or

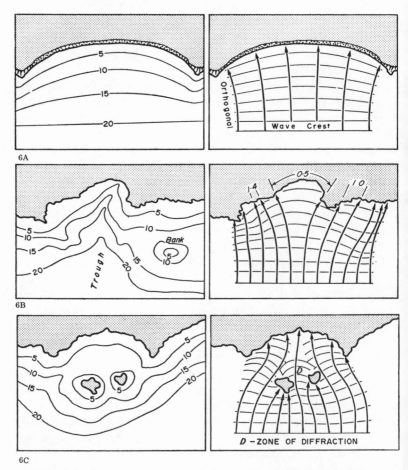

6A *Wave refraction pattern in an embayment of simple configuration (depths in fathoms)*

B *Wave refraction pattern in an embayment with a trough offshore and a submerged bank (some refraction coefficients are indicated)*

C *Wave refraction and diffraction patterns formed where a coast is bordered by islands* entrances are modified by diffraction, spreading out in the water beyond (Fig. 6C).

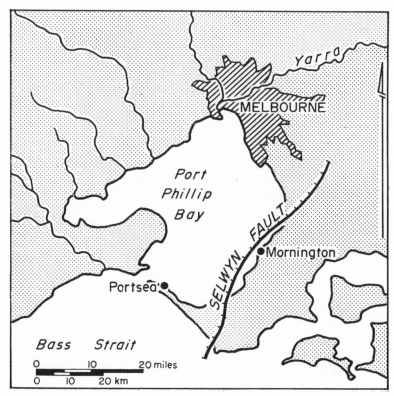

7   *Port Phillip Bay, Victoria*

Patterns of wave refraction in coastal waters can be traced from vertical air photographs, but methods have been devised for determining them graphically (Johnson *et al.*, 1948). Given basic knowledge of the direction of approach of waves in deep water, their length or period, and the detailed configuration of the sea floor, it is possible to construct diagrams predicting the patterns of refraction that will develop as they approach the coast. The modifications are best shown by orthogonal lines, drawn at equal intervals perpendicular to the alignment of waves in deep water, which converge where they pass over a submerged bank or reef, and diverge where deeper water is traversed; in general, they converge towards headlands and diverge in embayments (Arthur *et al.*, 1952). Orthogonals indicate how the wave energy is distributed, convergence towards a section of coast indicating a concentration of wave energy, whereas divergence of orthogonals indicates a weakening. The change in wave heights as waves are refracted is inversely proportional to the square root of the relative spacing of adjacent orthogonals. A refraction coefficient (R) can be derived as follows:

$$R = \sqrt{\frac{S_O}{S}}$$

where S_O is the distance between a pair of orthogonals in deep water and S their spacing on arrival at the shoreline. Calculated from a wave refraction diagram, this coefficient is an approximate expression of the relative wave energy on sectors of the coastline.

Diagrams of this kind have been used to examine the effects of waves on the profile of sandy beaches (Munk and Traylor, 1947), and it has been found that the position of such features as stream outlets or lagoon entrances are related to wave energy distribution as indicated by the spacing of orthogonals (Bascom, 1954). The outline in plan of many beaches (Plate 1) is related to patterns of approaching waves (Davies, 1959), and diagrams of wave refraction can also be used to predict the direction of longshore currents that develop where waves arrive at an angle to the shore (Shepard and Inman, 1950). Longshore currents also disperse from sectors of coast where wave energy has been concentrated by refraction.

The direction, height, and frequency of waves approaching a shore are of great importance in coastal studies. These parameters are subject to much variation, and it is usually necessary to make a statistical analysis of several years' records to establish characteristic annual and seasonal patterns (Helle, 1958). Direct observations are made from lightships and from certain coastal stations, but it is generally necessary to supplement these by analyses of meteorological data. On some coasts a particular wave direction is clearly dominant. The S coast of England has dominant waves from the direction of the prevailing SW winds, and the New South Wales coast receives a dominant SE swell. There is greater variation on coasts sheltered from regionally prevailing wind-wave and ocean swell patterns, such as the N Norfolk coast, where the prevailing SW winds blow offshore, and waves arrive from the NE, the N,

1   *Curved beach outline related to refracted pattern of ocean swell in Frederick Henry Bay, Tasmania*

2   *The blowhole cave at Lochard, near Port Campbell, Victoria*

or the NW according to local wind conditions. Records of onshore winds at Blakeney Point indicate that NE winds were dominant in 1958, and NW winds in 1960. Long-term records are necessary to determine dominant wave directions in these conditions.

Attempts have been made to correlate regional meteorological records with nearshore wave conditions. Silvester (1956) found that depression centres in one area off Western Australia produced waves approaching the coast from the NW, whereas depression centres in another area yielded southwesterly waves. Analysis of meteorological charts over a ten-year period gave the duration of depression centres in each area, and thus the relative incidence of northwesterly and southwesterly waves at the coast. Darbyshire (1961) outlined methods of predicting wave characteristics over the N Atlantic from synoptic records of wind velocity and direction applicable to 200-mile (320 km) grid squares in the ocean area. Characteristics of ocean swell spreading beyond a generating storm area were computed from diagrams of the angular distribution of wave energy, and added to the wind-wave pattern in each grid square. If wave refraction is taken into account, it is possible to predict wave conditions in coastal waters on sectors of the N Atlantic coast. Meteorological records can be analysed to indicate the relative duration of various kinds of wave régime over long periods, and to calculate resultants applicable to the study of process and change on coastal landforms. Few ocean areas are as well served by synoptic meteorological observations from ships, aircraft, and weather stations as the N Atlantic, but recent developments in satellite photography may produce useful data on wave patterns in less frequented ocean areas.

Coasts exposed to ocean swell and stormy seas are known as high wave energy coasts. Those that are sheltered from strong wave action are termed low wave energy coasts, and it is useful to recognise an intermediate category of moderate wave energy coasts (Davies, 1964). The stormy Atlantic coast of Britain and the ocean coasts of South Australia, W Victoria, and W Tasmania are high wave energy coasts subject to frequent strong wave action, with rugged cliffs and bold sweeping beach outlines. At the other extreme are low wave energy coasts bordering narrow straits, landlocked embayments, and island or reef-fringed seas where limited fetch or intense refraction prevent the development of strong wave action. Examples are found in the Danish archipelago, in some of the embayments sheltered by the Great Barrier Reefs on the Queensland coast, and on the shores of many estuaries and lagoons. The Florida sector of the Gulf Coast of the United States has wave action reduced by a broad, gently-shelving offshore profile (Tanner *et al.*, 1963) and shows features typical of a low wave energy coast: limited beach development, lack of cliffing, an intricate shore configuration with minor spits, deltas, and coastal swamps, and persistence of offshore shoals. The distinction should be made in terms of normal prevailing wave régimes, for this kind of coast is occasionally subject to drastic modification by strong wave action during hurricanes. Variations in tide range introduce a complication, wave energy being more effective on microtidal than on macrotidal coasts, where it is dissipated over a broad shore zone; wave energy also shows marked variations with aspect on coasts of irregular outline. Many coasts fall into the intermediate category, being partly cliffed, but having such features as spits, deltas, and swamps which diversify the outline of depositional sectors. Much of the E coast of England can be classified in this way and similar conditions obtain in gulfs and landlocked embayments around the Australian coast.

Currents

Ocean currents are drifts of water in response to prevailing wind patterns, and are also due, in part, to density variations in the oceans resulting from differences in the salinity and temperature of water masses. There is an analogy here with air masses in the atmosphere, where density variations resulting from differences in humidity and temperature produce atmospheric currents (i.e. winds). These ocean currents should not be confused with currents associated with waves and tides, or currents that develop locally in coastal waters when strong winds are blowing. True ocean currents are slow movements of water, and are not important in coastal geomorphology, except where they bring in warmer or colder water, which affects ecological conditions near the coast and thereby influences the world distribution of such landforms as coral reefs and mangrove swamps.

Tidal currents, produced by the ebb and flow of tides, alternate in direction in coastal waters, reversing as the tide ebbs; their effects may thus be temporary or cyclic. In the open ocean tidal currents rarely exceed 3·2 km/hr, but where the flow is channelled through gulfs, straits between islands, or entrances to estuaries and lagoons, these currents are strengthened, and may locally and temporarily exceed 16 km/hr. The velocity of tidal currents constricted in this way is related to tide range, the strongest currents being generated in macrotidal areas at spring tides. In the Raz Blanchart, between Alderney and Cap de la Hague in NW France, the tidal current attains about 16 km/hr under these conditions. Off the N coast of Australia, another macrotidal area, powerful currents develop in the straits bordering Melville Island, particularly in Apsley Strait, between Melville and Bathurst Islands. Tidal oscillations impinging on a coastline may set up longshore currents, as on the N Norfolk coast, where the longshore flow is westward during the two to three hours preceding high tide, then eastward for another two to three hours as the ebb sets in. These currents are of limited importance in terms of erosion, deposition, or sediment flow on the coast, their effects being subordinate to the effects of waves, and the associated currents developed by wave action in the nearshore zone.

Wave action is generally accompanied by wind-driven currents produced where winds cause mass movement of surface water, building up sea level to leeward and lowering it to windward. Hydraulic currents then disperse water from areas where it has been piled up, until normal levels are restored. Strong currents are produced when winds drive surface water into gulfs or straits, or into and out of estuary and lagoon entrances. These may strengthen the currents produced by tides in similar situations, and it is often difficult to separate the effects of the two. Wind-generated currents are not as regularly alternating as tidal currents, however, and their effects are likely to be cumulative in the direction of the prevailing wind. King (1953) has shown that winds blowing offshore can set up a shoreward bottom current in the nearshore zone.

Currents produced by fluvial discharge carry sediment into the sea, maintain or enlarge river outlets, and form a seaward 'jet' which refracts approaching waves and acts as a 'breakwater' impeding or interrupting longshore current flow. Discharge currents are strongest off streams fed by melting ice and snow from coastal mountains, as in Norway and Alaska during the summer months. In tidal estuaries, fluvial discharge is reduced, and possibly halted, by the rising tide, but augmented by the ebb.

Currents generated by wind and tide can be strong enough to move superficial sediment on the sea floor to prevent deposition and even to erode channels or colks (scour-holes) where water is driven through narrow straits such as the entrance to Port Phillip Bay (Fig. 8) (Benson, 1963). The sea floor between islands in Bass

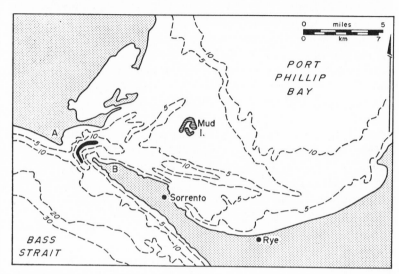

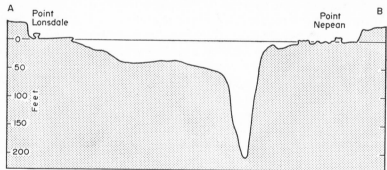

8   *Tidal colk at the entrance to Port Phillip Bay*

Strait shows deep, elongated channels, evidently formed by current scour (Jennings, 1959), and narrow entrances to lagoons and estuaries have often been deepened by the action of tidal currents (Fig. 50, p. 133). Closed basins have been scoured in soft outcrops on the sea floor off the Dorset coast and in the Bristol Channel by tidal currents (Donovan and Stride, 1961), and off the Brittany coast paths of current flow are marked out by rocky or gravelly sectors on a sea floor that is generally sandy or muddy. Shoals of gravel sometimes develop adjacent to rocky headlands in sectors of current flow, examples being the Skerries, E of Start Point, and the Shambles, E of Portland Bill, on England's S coast (Fig. 28, p. 106).

The effects of currents in transporting sediment were measured by the Swedish scientist, Hjulström, who showed that particles of sand (grain diameters between $\frac{1}{16}$th mm and 2 mm) are moved by currents of velocity greater than 20 cms/sec, but higher velocities are required to move larger and heavier particles, and also smaller particles, such as silt and clay, which are less easily moved because they are cohesive. The most readily eroded particles are medium sand ($\frac{1}{4}$ mm to $\frac{1}{2}$ mm), which are moved by currents of about 15 cm/sec (Hjulström, 1939). In the nearshore zone, sediment is more readily transported by relatively weak currents, because particles are thrown into suspension by the water turbulence associated with breaking waves.

Currents are generally more effective in shaping sea floor topography than in developing coastal configuration, and the early view that long, gently-curving beaches were produced by currents sweeping along the shore has given place to the modern view that these outlines are determined by refracted wave patterns (see Chapter V). Nevertheless, changes in the topography of the sea floor, due to erosion by current scour or deposition from slackening currents, modify patterns of wave refraction and may thus indirectly affect coastal outlines. Currents often play a part in removing material eroded by waves from the coast, or in supplying the sediment that is built into depositional forms by wave action. In addition, currents are responsible for the development of various kinds of ripple patterns on foreshores and tidal flats, on the floors of estuaries and lagoons, and on the sea bed. These points will be considered further in later chapters.

It is difficult to trace and measure currents accurately. Superficial movements of coastal water can be traced by introducing fluorescein, a coloured dye which indicates where the water is moving. Another technique is to use floats, submerged as far as possible to reduce wind resistance and carrying distinctive markers that can be observed, enabling their speed to be measured as they move with the current. It is often simpler to measure the effects of nearshore processes (volumes of sediment eroded, in transit, and deposited) than the processes themselves—a theme that will be taken up again in Chapter V.

Tsunamis

Exceptional disturbances of ocean water occur during and after earthquakes, landslides, or volcanic eruptions in and around the ocean basins. These produce waves several hundred miles in length, with periods of up to half an hour, proceeding at velocities of up to 800 km/hr across the deep oceans. They are barely perceptible in mid-ocean, but on entering shallow coastal waters they build up into giant 'tidal waves' which may attain heights of more than 30 m by the time they reach the coast. The term 'tidal waves' is misleading, for they are not tidal in origin, and the scientific term for them is the Japanese word 'tsunami'. They are most common in the Pacific Ocean, which is bordered by zones of crustal instability, and they are responsible for occasional catastrophic flooding and erosion of Pacific coasts, often with much devastation and loss of life far from the originating disturbance. In April 1946 a tsunami was initiated by an earthquake off the Aleutian Islands, and waves travelling southwards arrived in the Hawaiian Islands, 3700 km away, in less than five hours, having moved at an average speed of 750 km/hr. Shepard, who happened to be in Oahu at the time, has described how, after a preliminary withdrawal of water, a succession of giant waves broke at intervals of 12 minutes, each higher than the last, sweeping across the coral reefs and rising up to 10 m above normal sea level. The town of Hilo in Hawaii was hit by waves rising 9 m, and at one point a wave reached 16·8 m. Beaches were swept away, and the waves carried reef debris, including large blocks of coral, on to the coast, and eroded hollows on hillsides far above normal high tide level. Giant waves were recorded at many other places around the Pacific Ocean, and at Scotch Cap, Alaska, they destroyed a lighthouse and a radio mast 30 m above sea level (Shepard, 1963).

After these events, a network of seismographic and tidal observatories was set up by the U.S. Coast and Geodetic Survey in the Pacific region to record the initiation and give warning of the spread of tsunamis. The next major disturbance was an earthquake of exceptional intensity off the coast of S Chile on 22 May 1960, when huge waves broke on the adjacent coast, and very long waves spread radially across the Pacific Ocean. Soon afterwards, abnormally high waves were recorded on tide gauges on the Pacific coast of North America, and it was forecast that giant waves would reach Honolulu. The waves arrived at the predicted time, but they were only 1·2 m above normal; Hilo was again devastated, this time by waves reaching 11 m, and a few hours later waves up to 40 m high hit the coasts of Hokkaido and Honshu in Japan, causing extensive damage and drowning many people. Waves 3·3 m above normal were recorded at Port Lyttleton, on the E coast of South Island, New Zealand, and the tide gauge at Cronulla, on the coast of New South Wales, recorded a series of exceptional oscillations of water level up to 0·75 m above and below the predicted tidal curve. The Alaskan earthquake of March 1964 initiated a tsunami in the Pacific, and exceptionally high waves were recorded on the W coast of North America. The effects were moderate, however, in Hawaii, on the Japanese coast, and in the S Pacific; Sydney recorded tides about a foot higher than usual.

It is now realised that tsunamis are not necessarily 'damped down' by distance; the magnitude of waves received depends partly on offshore topography, the waves being higher where the offshore zone is gently shelving, and partly on the orientation of a coast in relation to the source of the disturbance. An earthquake along a fault line is likely to produce higher waves on coasts facing and parallel to the fault than on coasts which run obliquely to it; the greatest effects of the 1960 tsunami were on parts of the Japanese coast parallel to the line of disturbance of the Chilean earthquake. Wave heights are much reduced where coral reefs border the coast, where there is deep water close inshore, or where the waves have been refracted round reefs, shoals, or islands of intricate configuration.

Huge waves can be generated in restricted areas by landslides and rockfalls. Miller (1960) gave details of the effects of a massive rockfall into Lituya Bay, an Alaskan fiord, in July 1960. Water swept 520 m up an adjacent mountainside, and a wave 15 m high raced down the fiord at 160 km/hr, sweeping across a spit at the entrance and dispersing in the sea area offshore. Disturbances of a similar kind develop in the vicinity of ice coasts as the result of iceberg calving.

These are unusual features. On most coasts the tides, waves, and currents operate, and influence the development of coastal landforms, within a zone above and below present mean sea level, the extent of the zone depending largely on tide range. Storm waves achieve erosion and deposition several feet above high tide mark, and the sea floor is subject to the effects of wave action, increasing in intensity shorewards from the line where water depth equals half the wave length. It is clear that, in the past, the zone of operation of marine processes has been at times higher than it is now, and also, at times, lower; there have been 'stillstands' when the relative levels of land and sea remained constant at particular altitudes, and episodes when the levels of land and sea have changed. The evolution of coastal landforms has been much influenced by the sequence of stillstands and changing levels of land and sea, and it is necessary now to examine the evidence for these, to show how the history of such changes is being established, and to consider for how long marine processes have been operative with land and sea at their present relative level.

III

CHANGING LEVELS OF
LAND AND SEA

At various places on the coast there are stranded beach deposits, marine shell beds, and platforms backed by steep cliff-like slopes, all marking former shorelines that now stand above high tide level and beyond the reach of the sea. These are *emerged* shorelines, and they owe their present position either to uplift of the land, or a fall in sea level, or some combination of movements of land and sea that has left the coast higher, relative to sea level, than it was before. In addition, there are *submerged* coastal features, such as the drowned mouths of river valleys, submerged dune topography, and the former shorelines marked by breaks of slope that have been detected by soundings on the sea floor. Evidence of submergence has also been obtained from borings in coastal plains and deltas where freshwater peat and relics of land vegetation have been encountered well below present sea level, together with shoreline deposits formed when the sea stood at a lower level relative to the land. Submerged shorelines may have been caused by subsidence of the land, sea level rise, or a combination of these movements. Where evidence of a change in the relative levels of land and sea is found, it is often difficult to decide whether it is a result of uplift or depression of the land *(tectonic movement)*, raising or lowering of sea level *(eustatic movement)*, or a combination of the two. Until this is known, it is best to use the neutral terms *emergence* and *submergence*.

Some parts of the coast show evidence of both emergence and submergence, the sea having been higher relative to the land at one stage, lower at another, and now in an intermediate position. A close investigation is necessary to determine the sequence and dimensions of successive changes in the levels of land and sea, and whether any changes are still in progress. Some coasts are now being submerged, others are emerging. If no changes of land or sea level are taking place, a condition of 'stillstand' has been attained. The various kinds of tectonic and eustatic movement will be considered first, before dealing with the evidence of such changes on coasts in more detail.

Movements of the land

Upward or downward movements of the land, termed *tectonic* movements, may be *epeirogenic, orogenic,* or *isostatic.* Epeirogenic movements are broad-scale elevations or depressions of continents and ocean basins, with warping restricted to a marginal hinge-line, whereas orogenic movements involve more complicated deformation, with folding, faulting, warping, and tilting of the land. Broad-scale epeirogenic movements took place in Australia late in Tertiary times, when the Eastern Highlands and the large Precambrian shield region of Western Australia were elevated, while the intervening tract from the Gulf of Carpentaria southwards to the mouth of the Murray remained at a comparatively low level. Episodes of epeirogenic uplift have marked the evolution of the African continent, where broad planation surfaces initiated during intervening phases of tectonic stability are separated by bold scarps initiated during each uplift episode (King, 1962). Tectonic deformation at the margin of an epeirogenically-uplifted continent would be likely to run parallel to the general trend of the coastline, with elevation on the landward side and subsidence on the seaward side of a 'hinge line'. Bourcart (1952) considered this 'continental flexure' responsible for the stairways of terraces in which the oldest is the highest and farthest inland, reported from many coasts, but other possible explanations of these terrace sequences must be considered (see below, p. 38).

Orogenesis is the process by which mountains are built; it begins with the formation of vast crustal depressions, known as geosynclines, which subside as sediment is deposited in them, and are subsequently upheaved as chains of mountains. Orogenic movements are in progress N of Australia, in New Guinea and Indonesia, where active deformation is accompanying the formation of the geosyncline that lies between the more stable crustal regions of the Sahul Shelf, bordering N Australia, and the Sunda Shelf beneath the China Sea. In Victoria, Port Phillip Bay (Fig. 7) occupies a tectonic depression, or *sunkland,* bordered on the eastern side by Selwyn Fault, an active fault along which earthquakes still occur from time to time: the Mornington earthquake of 1932 was traced to a displacement along this fault. Earth movements continuing into Recent times have also influenced the levels of coastal features in New Zealand, Japan, California, and around the

Coasts

Mediterranean Sea. On the north coast of New Guinea near Aitape, Hossfeld (1965) obtained radiocarbon dates indicating that an emerged mangrove swamp had been uplifted about 50 m during the past five thousand years. Where warping has been transverse to the coastline, as in the Wellington district, New Zealand, there is a juxtaposition of uplifted and emerged coasts with downwarped and submerged coasts, and parts of the coastline have been determined by Recent faulting (Cotton, 1942).

Isostatic movements are adjustments in the earth's crust where it is heavily loaded by ice, lava, or accumulating sedimentary deposits. Crustal subsidence is taking place in the vicinity of large deltas, where the load consists of sedimentary deposits accumulating at the mouth of a river, and borings have shown that the Mississippi delta is underlain by a great thickness of Quaternary sediment, occupying a crustal depression, the subsidence of which is partly due to isostatic adjustments of the earth's crust beneath the accumulating sedimentary load. A long history of subsidence can be deduced from this stratigraphic evidence, for the deposits on river terraces bordering the Mississippi valley can be traced down into the wedge of sediment beneath the delta (Fig. 83, p. 186) (Russell and Russell, 1939). Other large deltas show similar evidence of subsidence as sediment accumulated.

Complicated isostatic movements have taken place in regions that were glaciated during Pleistocene times. As ice accumulated the earth's crust was depressed; when it melted, the crust was gradually elevated. In Scandinavia and N Canada the land is still rising because of the melting of glaciers and ice sheets. Isostatic recovery of depressed crustal areas continues for some time after the ice has gone, until the physical equilibrium of the earth's crust is restored, so that coasts bordering recently deglaciated areas have been subject to continuing uplift of the land; they show sequences of differentially elevated shorelines, as on the southern shores of Hudson Bay in N Canada and on the coasts of highland Britain (Stephens and Synge, 1966; Walton, 1966). Isostatic recovery has raised shoreline deposits 10,000 years old by as much as 275 m in the NW of the Gulf of Bothnia.

Each of these kinds of land movement may have affected, or may be affecting, the evolution of coastal landforms. Coasts bordering land areas that have been stable during Quaternary times show evidence of successive episodes of submergence and emergence which must be the result of upward or downward movements of the sea; it is only on such coasts that the record of Quaternary eustatic oscillations of sea level can be established. Unfortunately, it is difficult to prove that a section of coast has been stable during Quaternary times: it becomes a matter of increasing probability in regions without evidence of earthquakes or volcanic action, where the coastal rocks are relatively old, and where emerged and sub-merged shoreline features, traced along the coast, are found to be horizontal, above and below present sea level.

Movements of the sea

Upward and downward movements of sea level, termed *eustatic* movements, are world-wide because the oceans are interconnected. Major changes of sea level are caused by the addition or subtraction of water from the oceans, changes in the configuration and capacity of the ocean basins, or a combination of these two factors (Fair-bridge, 1961).

It is believed that the total quantity of water on and around the earth has been more or less uniform throughout geological time, apart from relatively small accessions of 'juvenile' water, supplied from the interior by volcanic processes. This means that water lost from the oceans must be gained either by the atmosphere or by the land. If the volume of water in the oceans remains constant, sea level changes may result from variations in the density of ocean water accompanying changes in temperature: it has been calculated that a fall of 1°C in the mean temperature of the oceans would contract their volume so that sea level would fall about 2 m. Estimates of Pleistocene variations of mean ocean temperature based on palaeotemperature measurements on fossils from ocean floor deposits (see below, p. 32) are within 5°C of the present temperature, so that this process could only account for sea level oscillations of about 10 m. Geomorphological evidence indicates that Pleistocene sea level oscillations amounted to more than 100 m, and so other mechanisms must be involved.

Sea level would rise if water were added to the oceans, and fall if it were removed. This is evidently what happened during Pleistocene times, when world-wide climatic changes produced a series of at least four main cold (glacial) phases separated by warmer (interglacial) phases, and followed by a warmer (post-glacial) phase in Recent times; subsidiary mild (interstadial) phases occurred during each glacial phase. As the climate became cooler, with the onset of each glacial phase, the growing glaciers and ice sheets were nourished by precipitation from the atmosphere, derived largely from evaporation over the oceans, but deposited and retained as snow and ice on the land. As a result, water was removed from the oceans, their volume diminished, and a world-wide lowering of sea level took place. When the glaciers and ice sheets melted, water was returned to the oceans, and world-wide marine transgressions carried the sea to relatively high levels. It has been calculated that if all the ice remaining on the earth melted, sea level would rise about 60 m, but at the maximum of the Pleistocene glaciation sea level was more than 100 m lower than it is now (Donn *et al.*, 1962). These large, world-wide movements of sea level during Quaternary times are termed *glacio-eustatic* oscillations, and various attempts have been made to establish their sequence and provide a time-table for the climatic changes which led to them.

Such changes are believed to be due to variations in solar radiation, and attempts have been made to project estimates of variations in solar radiation, based on astronomical theory, backwards through time. Graphs of hypothetical variations, compiled mathematically, show a sequence of warmer and colder periods during Quaternary times, and these have been correlated with interglacial and glacial phases. The time-scale of climatic oscillations thus deduced has been used by Zeuner (1959), Fairbridge (1961), and others for the dating of Pleistocene and Recent events, including sea level changes (Fig. 9). The duration of the Pleistocene was then estimated at 600,000 to one million years, but subsequently radiometric dating has suggested a longer duration, of the order of 2–3 million years.

An alternative approach uses oxygen isotope ratios measured in analyses of fossil foraminifera taken from sedimentary cores obtained from the floors of the oceans, where sedimentation has been very slow, and deposits marking the whole of the Quaternary era are only a few feet thick. The ratio of the oxygen isotopes O^{16} and O^{18} in fossil foraminifera is an indication of the temperature of the environment at the time they formed, and thus, compared with present ocean temperatures, an indication of warmer or colder climates in the past. Foraminifera obtained at successive levels from the stratified sequence of ocean floor sediments yield evidence of past changes of climate, with warmer and colder phases that

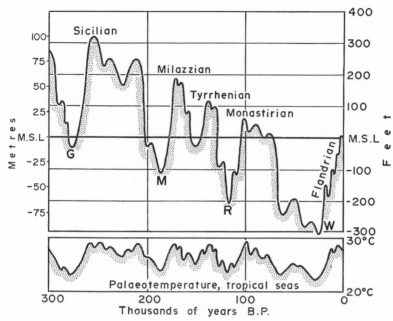

9   *Eustatic oscillations of sea level during Quaternary times (after Fairbridge, 1961). The peaks in the graph mark interglacial phase high sea levels, the troughs glacial phase low sea levels (G = Gunz, M = Mindel, R = Riss, W = Würm, the four European glacial phases); the lower graph shows Quaternary temperature variations in tropical seas. Zeuner (1959) gave a longer time-scale for these events, placing the Sicilian at 660,000, the Milazzian at 500,000, the Tyrrhenian at 270,000, and the Monastirian at 150,000 years B.P.*

are correlated with interglacial and glacial phases in the Pleistocene period; these have been graphed by Emiliani (1963), and the graph is appended to Fig. 9. It is supposed that each climatic oscillation was matched by a glacio-eustatic oscillation of sea level, a warmer ocean indicating a higher sea level, and a cooler ocean a lower sea level. In other words, the temperature graph will do for a graph of changing sea levels if it can be calibrated with vertical dimensions. If the minimum temperature is regarded as equivalent to the maximum lowering of sea level by more than 100 m, this provides a scale which permits the magnitude of successive glacio-eustatic oscillations to be gauged. The problems of relating this graph to the geomorphological evidence of higher and lower sea levels on coasts that have been stable in Quaternary times will be discussed later.

Sea level changes may also result from variations in the capacity of the ocean basins, the volume of ocean water remaining constant. If the capacity of an ocean basin (measured below present sea level) were increased, sea level would fall; if it were reduced, sea level would rise. It should be noted that we are now no longer dealing purely with movements of sea level, for changes in the configuration and capacity of the ocean basins involve tectonic movements in the earth's crust beneath the oceans or on bordering coasts. Epeirogenic sinking of an ocean basin would increase its capacity, and cause a world-wide lowering of sea level, while orogenic or isostatic movements in and around the ocean basins would lead to a world-wide lowering or raising of sea level, according to whether they increase or diminish the capacity of the ocean basin. As tectonic movements anywhere within the ocean basins will influence sea levels all over the world, it is best to distinguish sea level changes of this kind as *tectono-eustatic*.

Deposition of land-derived sediment in the sea and outpouring of volcanic lava on the sea floor also reduce the capacity of the ocean basins and therefore raise sea level, but these modifications would not lead to sea level changes as large or as rapid as those due to glacio-eustatic or tectono-eustatic variations. According to King (1959), transference of all the land above present sea level into the ocean basins would raise the level of the oceans by more than 250 m, but present estimates of denudation rates would only account for a secular rise of sea level of about 3 mm per century.

The distinction between glacio-eustatic movements of sea level, from changes in water volume while the capacity of the ocean basins remains constant, and tectono-eustatic movements of sea level, from changes in the capacity of the ocean basins while the water volume remains constant, is an academic one in terms of what is known of the geophysical history of the earth, for there is evidence that the volume of the oceans and capacity of their basins have both changed during the upward and downward movements of sea level through Quaternary times. In practice, eustatic movements are the world-wide equivalent changes of sea level recorded by geomorphological evidence on coasts where emerged and submerged features are not caused by movements of the land. On coasts that have not been stable, emerged and submerged features are the outcome of upward and downward movements of the land accompanying the world-wide eustatic movements of sea level.

The task of distinguishing between local land movement changes and world-wide sea movement changes is a difficult one, and will remain so until the record of world-wide eustatic movements has been firmly established on coasts where stability through Quaternary times can be proved. Meanwhile it is useful to collect evidence of the sequence of relative changes of land and sea level on particular coasts during Quaternary times, for it should eventually be possible to isolate the effects of eustatic oscillations of sea level, and thus to distinguish the effects of upward and downward movements of the land.

Measuring changes of level
Where emerged or submerged shoreline features are discovered, it is necessary to determine their levels above or below present sea level. Confusion has been caused by the use of different datum levels in surveys of this kind. Some workers have measured from mean sea level, the datum used on topographic maps in many countries; others from low water spring tide level, the datum used on nautical charts of coastal regions; and others from high water spring tide level, which is more easily determined and more readily accessible for surveying work on the coast.* Each of these has advantages for particular purposes, the first fitting in with surveyed contours and bench marks on topographic maps, the second enabling chart soundings and submarine contours to be used without modification, and the third being the most practical in field work, but the discrepancies between them increase with tide range, and it is necessary to adjust levels determined by different workers to the same datum before comparisons are made. Perhaps the best method would be to adjust all measurements of emerged and submerged shoreline features to mean sea level datum, stating the local spring tide range above and below this level.

Emerged shoreline features usually consist of benches or terraces, backed by abandoned cliffs, and sometimes bearing beach deposits or marine shell beds. Where possible, the level is measured at the back of the terrace, towards the base of the abandoned cliff, but this point is frequently obscured by beach deposits, dunes, or sediment washed down from adjacent slopes, and it is rarely

---

* High tide levels are indicated by strandline drift material, or the vertical extent of rock-dwelling marine organisms such as the rock oyster, which is generally found to an upper limit close to high water neap tide.

possible to determine it within $\pm$ 1·5 m. For this reason, the altitudes of emerged shoreline features are usually give in approximate figures: e.g. '5–8 m above mean sea level'.

Sequences of terraces measured on one sector of coast may not be recognisable on adjacent sectors, for terrace preservation requires a particular relationship between rock resistance, the intensity and duration of marine denudation at the higher shoreline phase, and the degree of subsequent degradation and dissection. As a rule, permeable sandstones, limestones (including coral), and gravel-strewn formations retain terraced forms better than impermeable rocks that are more resistant (restricted terrace development at the higher shoreline phase) or less resistant (greater destruction of terrace by subsequent denudation). Coastal terraces are rarely well preserved on rapidly-weathering rock formations.

Emerged shoreline features have often been dissected by valleys incised by rivers as they extended their courses across the emerged sea floor, whereas submerged shoreline features are more often preserved intact, and may be traced across the sea floor by sounding, providing they have not been obscured by later sedimentary deposition. Where fragments of emerged marine terraces occur at intervals along a coast, the problem of correlation arises. There are dangers in correlating terraces simply in terms of their altitude above present sea level, for where the land has been tilted laterally along the coast, fragments of the same terrace will be found at different levels, and fragments of different terraces may be found at similar levels. If there are sequences of terraces spaced at equivalent intervals above present sea level, correlation is less hazardous, and where this is possible the terraces indicate either a series of episodes of uniform uplift of the land without tilting or warping, or a succession of eustatic changes of sea level on the margins of a land area that has been stable.

Where marine deposits are associated with coastal terraces, correlation may be possible in terms of distinctive groups of fossils. Distinctive assemblages of minerals may occur in old beach sands, but the lateral variations in mineral composition of present-day beaches are such that correlation in terms of this evidence can only be tentative. A more promising means of correlation, developed in recent years, is based on radiocarbon measurements on samples of wood, peat, shells, or coral, obtained from deposits associated with former shorelines. The proportion of the carbon isotope C^{14}

to stable C^{12} diminishes with the age of organic materials, and is thus a measure of the time that has elapsed since they were alive. Radiocarbon analysis permits estimates of the age of samples less than 35,000 years old, and is therefore a means of distinguishing Recent deposits (less than 20,000 years old) from those formed during Pleistocene times. Dates obtained from radiocarbon analysis are stated in years B.P. (Before the Present), and there is an international agreement that the present should be defined as the year A.D. 1950. The margin of error is usually indicated as $\pm$ the standard deviation of the measurements made (e.g. 5,580 $\pm$ 200 years B.P.), which means that there is a 68·27 per cent chance that the true age lies within the range given, and a 95·45 per cent chance that it lies within *twice* the range given.

Certain difficulties in the application of radiocarbon dating have become evident in recent years. Radioactive contamination of the atmosphere by nuclear explosions has led to a sharp increase in the proportion of C^{14}, more than offsetting the dilution due to the combustion of fossil fuels, releasing 'ancient' carbon (i.e. with little or no C^{14}) into the atmosphere since the beginning of the industrial era. The C^{14} concentration in the atmosphere may also have varied in the past in relation to changes in the intensity of cosmic radiation, and to climatic variations. Dates obtained from shells or coral are less reliable than those from wood or peat because of the possibility that during their lifetime the calcareous organisms ingested an unknown proportion of ancient carbon derived from fossil organisms or dissolved carbonates, giving a deceptively high radiocarbon age. Alternatively, younger carbon from percolating groundwater may have replaced some of the fossil carbon, giving a deceptively low radiocarbon age. Carelessly collected or inadequately cleaned samples may also be contaminated by 'modern' carbon (e.g. if the roots of living plants have penetrated fossil wood or subsoil peat), traces of which can reduce the radiocarbon age to a fraction of the true age. It is hoped that other dating techniques will be developed which permit a more accurate dating and correlation of Pleistocene coastal deposits. Refinements of the potassium-argon method have recently been applied to the dating of volcanic materials interbedded in the Pleistocene stratigraphy of New Zealand, and it is likely that research of this kind will throw new light on the chronology of Pleistocene sea levels (Stipp *et al.*, 1967).

Evidence of emerged shorelines

Emerged marine terraces have been found on many coasts at various levels, and submerged terraces indicative of former shore-lines at lower levels have been discovered during surveys of sea floor morphology. Correlation of these former shorelines in terms of a sequence of Quaternary eustatic movements of sea level, allowing for the extent of tectonic movements in coastal regions during Quaternary times, has proved extremely difficult, and there is little prospect of an early solution to these problems. At one time it appeared that the sequence of terraces reported half a century ago by De Lamothe in Algeria and Depéret on the French Riviera, sectors of the Mediterranean coast thought to have been stable during Quaternary times, was being confirmed by studies of 'coastal stairways' elsewhere (Zeuner, 1959). The Mediterranean sequence consisted of four main high sea level stages (named Sicilian, Milazzian, Tyrrhenian, and Monastirian), taken as equi-valent to successive high stillstands of sea level during interglacial phases of the Pleistocene period. The Sicilian was equivalent to a sea level at 90–100 m (about 295–330 feet), the Milazzian 55–60 m (about 180–200 feet), the Tyrrhenian 28–32 m (about 90–105 feet), and the Monastirian was subdivided into Main Monastirian (18–20 m or 59–65 feet) and Late Monastirian (7–8 m or 23–26 feet). There was a tendency to correlate terraces discovered at these levels on other coasts with the Mediterranean sequence, and accept the correlation as evidence of the Quaternary stability of the coast concerned. In Britain, high-level marine deposits in SE England at about 200 m (650–700 feet) were assigned an early Pleistocene (Calabrian) age on palaeontological grounds, their contained fossil assemblage matching marine deposits of this age elsewhere. Terraces at successively lower levels were correlated with the Mediterranean sequence by Zeuner (1959), who postulated a high late Tertiary sea level followed by the glacio-eustatic oscilla-tions with successive interglacial maxima of diminishing altitude, culminating in the final low sea level of the Last Glaciation, and the succeeding Holocene marine transgression to present sea level. If all the ice remaining on the earth were now to melt, sea level would rise about 60 m so that if the sea stood about 200 m higher at the beginning of Pleistocene times something more is required than straightforward glacio-eustatic oscillations. One suggestion has been that the ocean basins, notably the Pacific Ocean basin, have been subsiding epeirogenically during Quaternary times, resulting in successively lower interglacial sea levels, so that the sea level fluctuations are partly tectono-eustatic changes. Another is that the Antarctic ice sheet became progressively larger, taking up an increasing volume of ocean water, during Pleistocene times.

In the last few years, doubts have arisen concerning the validity of the Quaternary sea level sequence outlined by Zeuner. On the one hand, it is possible that some of the classic sites around the Mediterranean, on which the original levels and chronology were based, were not tectonically stable during Quaternary times; some at least of the 'type areas' may owe their elevation, at least in part, to tectonic uplift (Castany and Ottmann, 1957). This question is being actively investigated and it would be premature to pronounce upon it here. On the other hand, it appears that altitudinal correlations of coastal terraces are not as good as was formerly thought. It has been all too easy to give emphasis to those terraces which appeared to lie within the range of levels quoted for the classic Mediterranean sequence or, admitting the possibilities of error in determining coastal levels, to agree that some of them were approximately equivalent to the Mediterranean sequence. When the results of detailed studies of coastal terraces are examined, they show considerable departures from the classic Mediterranean sequence of four or five main terraces up to and including the Sicilian (90–100 m): some workers have found many more terraces within this height range, others have found fewer, and former sea levels supposedly of Pleistocene age have been reported at almost every altitude below 200 m on various coasts.

The variety of viewpoints expressed in a symposium, 'Pacific Island Terraces, Eustatic?', held at Honolulu in 1959 (Russell, 1961) is symptomatic of the present situation. At one extreme are workers who accept most coasts as having been stable during Quaternary times, and envisage a sequence of eustatic movements of sea level with terraces marking former higher stillstands up to 200 m or more (Stearns, 1961) above present sea level. At the other extreme are those who are doubtful if Pleistocene sea levels ever stood more than 10 m (about 33 feet) above present sea level (Russell, 1964), and who would presumably argue that Quaternary coastal terraces and shorelines above this level owe their present elevation, at least in part, to tectonic uplift. These contradictory

views can only be resolved by carrying out more detailed studies
of the sequence of coastal terraces and associated stratigraphic
features, with former shorelines measured and dated as accurately
as possible, and reviewing the world situation. If coastal terraces
of Quaternary age up to 200 m above sea level are as widespread
as has been claimed, the former view requires large-scale movements of the ocean surface during Quaternary times, and the latter
view requires large-scale tectonic uplift over extensive coastal
areas during the same period. The discovery of similar marine
Pleistocene stratigraphic sequences at different coastal altitudes
indicates the possibility of unequal vertical displacement of former
shorelines (Flint, 1966). At this stage it will be useful to summarise
briefly the record of coastal terraces, on which an enormous
literature now exists (Richards and Fairbridge, 1965).

   In the British Isles, research on coastal terraces has been
dominated by the tracing of early Pleistocene shoreline deposits
210 m (690 feet) above sea level in SE England, and 'erosion surfaces' at a similar level in Wales. Beneath this level, fluvial terraces
in the river valleys descend towards former base levels estimated
at about 400, 200, 100, 50, 25, and 10 feet above present sea level.*
A correlative 'stairway' of coastal terraces has been recorded by
various workers along the S coast of England, and assigned a
Pleistocene age on the grounds that even the lower terrace features
and associated deposits are overlain by periglacial drift. The presence
of erratic boulders, presumably ice-rafted, on the Cornish coast
at and above present sea level poses a difficult problem, for it has
been assumed that climatic conditions cold enough for icebergs to
reach this latitude would have been accompanied by lowered sea
levels. If Zeuner's thesis of an overall lowering of sea level through
Pleistocene times is accepted, the possibility arises that these ice-
rafted erratics arrived during an early Pleistocene glacial phase
when the sea was lowered to about its present level (Stephens and
Synge, 1966). In E England coastal terrace levels have been
modified by subsidence in the North Sea basin, and in N and W
Britain the levels have been complicated by isostatic recovery
following deglaciation (Walton, 1966). The assumption that south-
ernmost England has been tectonically stable throughout Quaternary times, so that river and coastal terraces match former eustatic

---

* These estimates can be rendered, without much loss of real accuracy, as 120,
60, 15, 7·5, and 3 m respectively.

sea level stillstands, remains to be checked against worldwide evidence. Terrace sequences on the coasts of W Europe raise similar problems to those in S England.

In North America, the Atlantic coastal plain bears depositional terraces which have been correlated with successive alternations of sea level since Miocene times, the sea attaining levels of 13·5, 6, 7·5, 4·5, and 0 m respectively, with intervening low sea level phases (Oaks and Coch, 1963). On the Pacific coast, terraces have been dislocated by the San Andreas fault near San Francisco, and warped and uplifted terraces are found near Los Angeles and on Santa Catalina Island (Emery, 1960). In Chile, coastal terraces have been displaced tectonically in a region subject to earthquakes, and as the Chilean coast runs parallel to the main axes of Andean uplift there is a possibility that terraces found at the same height above sea level have been epeirogenically uplifted (Fuenzalida *et al.*, 1965).

Stairways of coastal terraces have been reported on the margins of continents assumed to have been stable in Quaternary times, as on the coast of India (Chaterjee, 1961), S Africa (Haughton, 1963), and S Australia (Gill, 1961), and in each case equivalence with former eustatic sea levels has been considered probable. The dangers and difficulties of terrace correlation are more obvious in regions of Quaternary tectonic activity, such as New Guinea, Japan, and New Zealand.

The evidence so far obtained from the Australian coast supports the idea of a sequence of successively lower interglacial sea levels, but much work remains to be done. There are certainly marine terraces at various levels, and related series of river terraces bordering valleys that descend towards the coast, but few detailed surveys have been made, and terrace levels have often been stated within wide limits, based on reconnaissance surveys, or merely on inspection of contoured maps of uncertain accuracy. Facile correlations of terraces in terms of approximate heights may well have concealed evidence of tectonic deformation in coastal regions, and favoured too strongly the view that these terraces correspond to former eustatic sea levels on coasts assumed to have been stable.

Marine terraces have been reported on various parts of the Australian coast at about 130 m, 40 m, 12–15 m, and 6–9 m above present mean sea level (David, 1950; Gill, 1961; Jennings, 1961). The last three are widespread, and could correspond with interglacial or interstadial eustatic stillstands of high sea level. Associated beach deposits have been found to contain marine shells of types that now occur in the warmer parts of the ocean, and oxygen isotope measurements have generally confirmed a warmer environment of origin. A Pleistocene age is suggested by the fact that associated deposits are beyond the range of radiocarbon dating (i.e. older than 35,000 years), and it is therefore likely that the beaches formed on the shores of warmer high-level interglacial seas, but there is as yet no means of deciding which interglacial they represent (Gill, 1961).

In the SE of South Australia a series of stranded beach ridges runs roughly parallel to the coast at successive levels up to more than 60 m above present sea level. It was once thought that these corresponded with the Mediterranean sequence of interglacial high sea levels, but it is now realised that they were tilted transversely when the Mount Gambier region rose and the Murray-mouth area subsided during Quaternary times. The levels of the successive beach ridges do not, therefore, correspond with Quaternary eustatic high sea levels, for their emergence results from land movements as well as sea level oscillations. The same is true of the NE coast of New Guinea where a remarkable series of emerged Quaternary coralline terraces gains elevation and multiplies across an axis of uplift in the vicinity of the Tewai gorge, where the highest attains about 750 m (2500 feet) above sea level.* Cotton (1951a) has described similar warped and faulted coastal terraces from the vicinity of Wellington, in New Zealand.

Submerged shorelines

The development of accurate sounding devices has led to a more precise knowledge of the submerged topography of the sea floor bordering coasts, and the search for oil, natural gas, and mineral deposits is rapidly increasing our awareness of the structure and morphology of continental shelves. Stairways of submerged terraces have been mapped off various coasts, one of the clearest being the paired sequence of seven terraces at successive levels down to 110 m bordering Tsugaru Strait, between Hokkaido and Honshu, in Japan (Emery, 1961). A submerged shoreline at a depth of about 18 m has been reported off the coasts of the United States (Shepard,

---

* J. M. A. Chappell, who is investigating these terraces, has found evidence of uplift 250 m in the past 30,000 years.

1963), and detailed studies have been made of beach deposits indicative of lower sea level stands on the floor of the Gulf of Mexico (Curray, 1960). Off S California the sea floor has been tectonically deformed (Emery, 1960). In some areas it is possible that smoothing of sea floor contours by wave action during episodes of marine transgression has obliterated or concealed evidence of submerged shorelines.

In Australian waters the classic evidence of a low sea level stage is the nearshore sediment dredged from a depth of 130 m off the New South Wales coast (Smith and Iredale, 1924). Jennings's investigation of the submarine topography of Bass Strait revealed a well-marked break of slope, indicative of a submerged shoreline at a depth of 60 m (Jennings, 1959). These features probably date from episodes of low sea level corresponding with glacial phases in Pleistocene times.

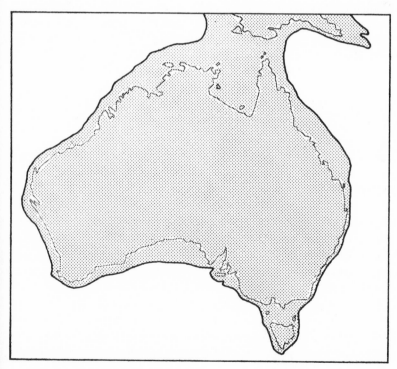

10   *The outline of Australia during the Last Glacial phase, when the sea stood about 140 m below its present level*

Lowering of sea level during the Last Glacial phase has been estimated from the volume of water taken up to form the calculated extent of glaciers and ice sheets during this phase. According to Donn *et al.* (1962) the sea stood between 110 and 130 m below its present level before the start of the succeeding marine transgression. Another estimate, based on extrapolation of the pre-Recent floor of the Mississippi valley out to the edge of the continental shelf, indicates a lowering of 140 m (Fisk and McFarlan, 1955). It is clear that a broad tract of the continental shelf was laid bare at this stage, and that the coastlines of Pleistocene interglacial phases were left high and dry, some distance inland. Britain was a peninsula of W Europe, and Australia, linked to Tasmania and New Guinea, had the outline shown in Fig. 10.

Holocene marine submergence

The subsequent transgression which carried the sea to its present general level and established the broad outlines of the world's existing coasts took place within the range of radiocarbon dating, and has been investigated from samples obtained from the sea floor, and from borings in deltas and coastal plains in various parts of the world. Valley mouths drowned by Holocene submergence contain the marine, estuarine, and deltaic deposits that accumulated during and since the phase of submergence. Carbonaceous materials found in these sediments can be dated by radiocarbon analysis and correlated with former sea levels. Dates have been obtained from samples of beach shells or shallow-water organisms, such as oysters, referable within narrow limits to sea level at the time of their formation, but more reliable ages are given by peat deposits and tree stumps found in borings and referable to levels slightly above the sea level at the time of their formation. In some cases these may now be below the level at which they formed, because of subsequent compaction of underlying peat or clay, or because of the isostatic subsidence that has taken place beneath many deltas. If attention is confined to coastal regions believed to have been stable during Recent times, radiocarbon dates plotted against sea level at the time of formation of these organic materials give a picture of an oscillating rise of sea level beginning about 20,000 years ago; in Europe this is known as the Flandrian transgression (Fig. 11).

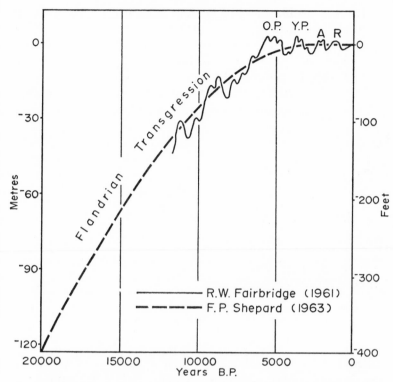

11   *The Holocene marine transgression. Alternative interpretations of sea level changes during the past 6000 years are indicated. Fairbridge considers that the sea rose to higher levels (O.P. = Older Peron, Y.P. = Younger Peron, A = Abrolhos, and R = Rottnest stages); Shepard thinks it attained present level without exceeding it during Recent times.*

The question of Recent emergence

It is widely agreed that a rapid transgression of the sea took place between 17,000 and 7000 years ago, but there is less agreement concerning changes during the past 7000 years. Shepard (1963) takes the view that sea level stood 6–9 m lower 7000 years ago than at present, and that during this period it has risen gradually to attain the present level: it may still be rising (see below, p. 47). Coleman and Smith (1964) consider that the sea attained its present level between 3000 and 5000 years ago, since when there has been a stillstand. Fairbridge (1961) believes there is evidence that the sea rose to higher levels during the past 6000 years, and that there have been three brief stillstands 3, 1·5 and 0·5 m above the

present level within this period (Fig. 11). Shepard and Coleman
and Smith base their views on the chronology of levels established
on the Gulf Coast of North America, and supported by measurements from many other parts of the world. Fairbridge bases his
view on the radiocarbon dates obtained from deposits associated
with coastal terraces of Western Australia (Fig. 12), and similar

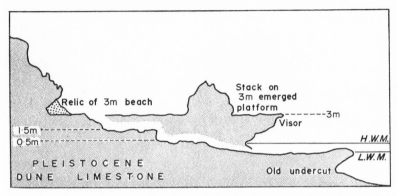

12   *Coastal platforms near Fremantle, Western Australia. Fairbridge (1961)*
correlates the three platforms with higher Holocene sea level stillstands.

dates for low terraces have been reported elsewhere around the
Indian and Pacific Oceans, on a scale suggestive of world-wide
eustatic oscillations rather than tectonic uplift. The differences
focus on the interpretation of coastal terraces and associated shoreline features suggestive of emergence found below the 6–9 m
shoreline of late Pleistocene interglacial times and generally within
3 m of present sea level. Russell (1963) obtained a radicoarbon age
of 35,000 ± 3700 years B.P. from coral fragments in a beach
associated with a coastal terrace at Carnarvon, Western Australia,
related to a 3 m higher sea level, and argued that the low terraces
on this coast pre-date the Last Glacial lowering of sea level; Hails
(1965) reached a similar conclusion in E Australia. The question
remains controversial. There are many emerged reefs and atolls in
the Pacific Ocean, but some of them have certainly been uplifted
tectonically, and even tilted (see p. 206), and these cannot be
accepted as evidence of a Recent higher sea level and subsequent
eustatic lowering.

Much more detailed work is needed to resolve this problem. It
is necessary to confine attention to features definitely indicative of

emergence, ruling out those that could have been formed by weathering processes, storm wave activity, or depositional processes with the sea at its present relative level. It is then necessary to find materials suitable for dating in a context that will give the age of the emerged feature, indicating whether it is Pleistocene or Recent. Emerged shore forms of Pleistocene age must have persisted through at least one phase of glacially-lowered sea level, when stream incision and dissection may have occurred. Emerged shore forms of Recent age could be indicative of a phase of higher stillstand of sea level if such features are indeed world-wide on coasts that have been tectonically stable since Pleistocene times. Tectonic uplift is likely to have been localised to particular sectors of coast, and variable in extent of vertical displacement, but the possibility of broad-scale, possibly continent-wide, epeirogenic uplift giving equivalent emergence over large areas cannot be ruled out.

Changes of level at the present time

Tide gauge records over the past few decades show evidence that changes in the relative levels of land and sea are still in progress. Mean sea level, determined by averaging high and low tide levels over a period of several years, is often found to have risen or fallen when compared with levels determined fifty or a hundred years ago (Gutenberg, 1941; Valentin, 1952). The changes thus recorded are again caused by movements of the land, movements of the sea, or combinations of both. The majority of tide gauges, including all those on coasts believed to have been tectonically stable in Recent times, show evidence of submergence during the past few decades, and as this evidence is widespread it is probable that a world-wide eustatic rise of sea level has been taking place, accompanying the melting of the margins of glaciers and ice sheets during this period. Ice recession is widely in evidence in the northern hemisphere, and the release of water to the oceans has evidently exceeded the loss incurred in the increase of Antarctic ice volumes over this period (Wexler, 1961). There appears to have been a slight warming of atmospheric temperatures, accentuated by the increasing proportion of carbon dioxide resulting from combustion of fossil fuels, particularly during the industrial era.

It has been calculated that the sea has risen at the rate of 1 mm per year, or about 4 inches during the past century. Tide gauges that show a more rapid rate of submergence than this indicate that the local coast has been sinking tectonically; those that show a smaller submergence, or an emergence, indicate that the land has been rising. Donn and Shaw (1963) have summarised the variations in tidal levels on the United States coast during the past century. In Britain, tide gauge records suggest uplift in the N and continuing subsidence on the E and S coasts. Repetition of precise geodetic levelling from time to time can provide an independent check on the results obtained from tide gauge records, and a means of determining areal patterns of crustal deformation more accurately than from measurements of tide levels at scattered coastal stations. Over the next century it should be possible to decide which indeed are the stable areas of the earth's crust, and to isolate more accurately the extent of eustatic sea level change.

A number of geomorphological effects have been attributed to the rise in sea level over the past century. They include the renewal of coral growth on planed-off reef surfaces off the coast of Western Australia (Fairbridge, 1947), the onset of erosion on sandy shores, notably in Australia (Chapter VI), and the prevalence of erosion on the seaward margins of many marshlands (Chapter VII). In each case it is possible to indicate other relevant factors and processes, but the continuance of a slow marine transgression should be borne in mind in considering the dynamics of existing coastal features.

IV

CLIFFED COASTS

Marine erosion of cliffed coasts takes place mainly during storms, and is achieved largely by wave action: the hydraulic pressure of impact and withdrawal, and the abrasive action of water laden with rock fragments (sand and shingle) hurled repeatedly at the cliff base. After a storm the backshore is littered with debris eroded from the cliff. Much of this becomes broken and worn down by wave action (a process known as attrition), and is either retained as a beach, or carried away along the shore or out to sea by the action of waves and currents. The simplest type of cliffed coast is found where marine erosion has attacked the margins of a stable land mass of homogeneous and relatively resistant rocks, removing a wedge of material to leave a steep cliff at the back of a gently-sloping platform, which extends from high tide level to beneath low tide level (Fig. 13A). There are, however, many variants of this simple form, for complications are introduced by the lithology and structure of coastal rock formations, the degree of exposure to wave attack, the effects of subaerial processes of denudation on the coast,

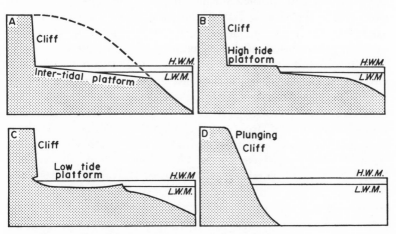

13A *Cliffed coast with an inter-tidal shore platform*
  B *Cliffed coast with a shore platform at about high tide level*
  C *Cliffed coast with a shore platform at about low tide level*
  D *Plunging cliff form, with no shore platform*

and the history of changing land and sea levels. The platform is often termed a wave-cut, or abrasion platform, but these generic terms can be misleading and the purely descriptive term, shore platform, is preferred here. A misconception repeated in many textbooks is that the platform extending beneath low tide level is continued by a wave-built terrace constructed by deposition at its outer margin. There is little evidence that such a terrace actually exists on sea coasts (Dietz, 1963).

The morphology of cliffs

Examination of almost any section of cliffed coast soon reveals features that are related to variations in lithology and structure, picked out by marine erosion. It is obvious that the more resistant parts of coastal rock formations protrude as headlands, or persist as rocky stacks and islands offshore, whereas the weaker elements are cut back as coves and embayments. Resistance in this context means the hardness of rocks attacked by the physical forces of marine erosion, or their durability in face of other processes at work on the coast, including the physical and chemical effects of repeated wetting and drying of rock surfaces, and the purely chemical effects of solution by sea water, notably on limestone coasts. Solid and massive formations are generally eroded more slowly than formations that disintegrate readily, such as friable sandstones, or rocks with closely-spaced joints and bedding-planes, or rock formations shattered by faulting. Weathering and marine erosion penetrate these lines of weakness, excavating caves and coves, so that patterns of jointing and faulting influence the outline in plan of a cliffed coast. Natural arches, formed where less resistant rock has been excavated from a promontory or islet, are exemplified by the Green Bridge of Wales on the S Wales coast which consists of massive Carboniferous Limestone beneath which thinly-bedded strata have been cut out along joint planes, and various 'London Bridges', notably at Torquay in Devon and Portsea on the Victorian coast in Australia. Where the forces produced by the hydraulic action of waves and the compression of trapped air puncture the roof of a cave, water and spray are driven up through blowholes, as on Porth Island, near Newquay in Cornwall, and on the Old Red Sandstone coast near Arbroath in eastern Scotland. An Australian example is shown in Plate 2. Spectacular steep-sided clefts are produced where the roofs of caves collapse, or

3   *The Twelve Apostles, stacks bordering the cliffed coast near Port Campbell, Victoria*

4   *The Needles, stacks in hard Upper Chalk at the W end of the Isle of Wight, England. The promontory is a ridge of steeply-dipping Chalk, with fresh cliffs on the side exposed to predominant SW waves (Scratchells Bay, right) and partly degraded cliffs on the more sheltered side, bordering Alum Bay (left). (Aerofilms Ltd)*

5 *Castellated cliffs in granite at Lands End, England (Aerofilms Ltd)*

6 *Cliffs of columnar dolerite near Cape Pillar in SE Tasmania (Australian Tourist Commission)*

where rock is excavated along lines of weakness at an angle to the shore. Examples can be found on many cliffed coasts: they are known as geos or yawns on the coast of Scotland, and zawns in Cornwall, and local place names commonly credit them with some diabolical function. Gorges of this kind are found on the bold sandstone cliffs of the Jervis Bay district in New South Wales, and near Port Campbell in Victoria, where powerful wave action has penetrated joints and bedding-planes to sculpture a variety of forms, including stacks and natural arches (Plate 3). Similar coastal topography is found on the red sandstones of the Devon coast, notably at Ladram Bay, and in northernmost Scotland, where huge cliffs of Old Red Sandstone face powerful wave action from the north Atlantic. The Old Man of Hoy is a spectacular stack in front of towering cliffs of Old Red Sandstone on the west coast of the Orkneys, but probably the best known stacks are the Needles, residual ridges of hard chalk at the W end of the Isle of Wight (Plate 4).

Certain kinds of lithology yield characteristic cliff forms. The castellated granite cliffs of the Lands End peninsula in Britain are related to cuboid jointing (Plate 5), and the columnar basalt cliffs of N Ireland and the similar columnar dolerites on the high cliffs of SE Tasmania (Plate 6) result from pronounced vertical jointing. In limestone areas the sea may penetrate and widen caves that originated as subterranean solution passages: the caves of Bonifacio, in S Corsica, are believed to have formed in this way.

Where the dip of coastal rock formations is seaward, undercutting by marine erosion often leads to landslips and rock falls, the undercut rock sliding down bedding-planes into the sea, leaving the exhumed bedding-planes as a coastal slope. On the S coast of England, coastal landslips are common where the permeable Chalk and Upper Greensand formations dip seaward, resting on impermeable Gault Clay. Water seeping through the permeable rocks moves down the clay plane, and if marine erosion has exposed the junction at or above sea level, lubrication of the interface leads to slipping of the overlying rocks. A spectacular landslip occurred in this situation on the E Devon coast near Axmouth in 1839 (Arber, 1940), and similar slides have occurred on the S side of the Isle of Wight, and between Folkstone and Dover, where a railway built along the undercliff is damaged from time to time by falling rock. It is probable that the physical effects of wetting and lubrica-

7   *Lulworth Cove, on the Dorset coast, S Englana, a cove excavated in Wealden sands and clays behind a breached wall of strongly-folded Jurassic limestone and backed by a high ridge of Chalk (Aerofilms Ltd)*

8   *Slope-over-wall cliff topography on the W side of Dodman Point, Cornwall, England. The promontory consists of intensely folded slates and phyllites, exposed in the lower part of the cliff, but mantled by periglacial rubble in the upper slope. (C. T. Bird)*

tion of the Gault Clay surface are accompanied by chemical processes, involving base exchange. The seeping water is rich in dissolved calcium carbonate, and when it reaches the glauconitic Gault Clay, calcium ions displace potassium ions from the clay, alkalinity increases, and the clay is deflocculated (Varnes, 1950). Conversion of the upper layers of Gault Clay into a soft wet slurry which flows out at the base of the cliff hastens the undermining of the Chalk and Upper Greensand beds. Cliffs on soft clay formations in Bournemouth Bay, and on parts of the East Anglian and Yorkshire coasts, are subject to recurrent slumping, particularly after wet weather. Subsequent removal of slumped material by waves and currents at the base of the cliff then rejuvenates the profile, preparing the way for further slumping, so that the cliffs recede as the result of alternating marine and subaerial effects.

Where relatively resistant coastal rock formations are backed by weak outcrops, penetration of the outer wall by marine erosion is followed by the excavation of coves and embayments. The classic example of this is on the Dorset coast E of Weymouth, where several stages can be seen. Stair Hole, near Lulworth, is an early stage, a narrow breach in the outer wall of steeply-dipping Jurassic limestone. Close by, Lulworth Cove has a wider entrance through the limestone wall, and an almost circular bay carved out of Cretaceous sands and clays, backed by a high ridge of Chalk (Plate 7). Farther E a much broader embayment has developed, opening up the clay lowland corridor in front of the Chalk ridge at Worbarrow Bay (Fig. 14).

Cliffs exposed to powerful wave action are often shaped entirely by marine erosion. This is true of the high wave energy coast near

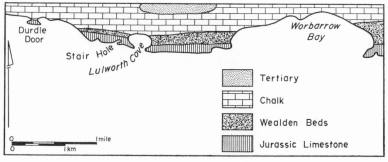

14 *Configuration of part of the Dorset coast, in S England.*

Port Campbell, in W Victoria, where steep cliffs have been cut by marine erosion in horizontal stratified Miocene sedimentary rocks, and the huge waves that break against these during storms have cut out ledges along the bedding-planes at various levels up to 60 m above high tide mark (Baker, 1958). The ledges are rarely more than 3–6 m wide, and it is important to note that they are the product of present-day storm wave erosion; they should not be confused with coastal terraces that bear 'raised beach' deposits indicative of emerged shorelines.

Cliffs on more sheltered sections of the coast, where strong wave action is intercepted by headlands, islands, or reefs offshore, or attenuated by a gentle offshore slope, may show features that have been formed by subaerial denudation as well as those shaped by marine attack. Cliffs in these situations often consist of a coastal slope shaped by rainwash and soil creep, the lower part of which is kept steep and fresh by wave attack. It is instructive to compare the bold profiles of cliffs of massive sandstone facing the ocean in the Sydney district with the gentler, often vegetated, slopes on the same geological formation on the sheltered shores of Sydney Harbour and Broken Bay. The degree of cliffing developed in these situations is closely related to the local fetch, which limits the strength of attack by local wind-generated waves.

Where beaches or barriers have been built up in front of a former cliffed coast, protecting it from marine erosion, subaerial processes become dominant and the steep sea cliff is 'degraded' to a coastal slope of gentler inclination, comparable with escarpment and valley-side slopes inland. In humid regions these slopes acquire a soil and vegetation cover. On the shores of Carmarthen Bay, in S Wales, growth of the Laugharne spit has cut off a former cliffed coast E of Pendine, and the subaerial evolution of slope forms on the abandoned cliffs has been studied by Savigear (1952). On the N Norfolk coast a line of bluffs, formerly cliffs, were cut off in a similar way by the development of spits, barrier islands, and marshlands (Fig. 30, p. 109). In this case the sea reached the cliffs in late Pleistocene times, the raised beach at Stiffkey dating from a late interglacial or interstadial phase, and deposition in front of them took place during and since the Holocene marine transgression, which reworked glacial drift deposits left behind on what is now the floor of the North Sea. In Australia, enclosure of former embayments by the growth of coastal barriers has been followed by the degradation of the former sea cliffs on the enclosed coast, as in the Gippsland Lakes region in Victoria (Bird, 1965).

The active cliffs that we now see on the coast have assumed their present form only since the Holocene marine transgression brought the sea to its present general level within the last 6000 years. Cliffed coasts undoubtedly existed before the Last Glacial phase of low sea level, but during that phase, in the absence of marine attack, they were degraded by subaerial denudation. In high latitudes this degradation took place partly under periglacial conditions, when the cliff became a slope mantled by frost-shattered debris. Since the Holocene transgression, periglaciated slopes of this kind have been undercut by marine erosion, but some coasts retain part of the periglaciated slope, the lower part having been cut back by wave attack to produce a 'slope-over-wall' profile (Plate 8). The slope, mantled by frost-shattered debris, is clearly a legacy of past periglacial conditions; it cannot be explained in terms of processes now at work. Coastal landforms of this type are well developed on the coasts of Devon and Cornwall, in SW England, where the proportion of relict periglaciated slope to actively-receding wall depends on the degree of exposure of a coastal sector to storm wave attack. On sheltered sectors of the S coast of Cornwall and Devon, the relict slope is well preserved, extending down almost to high tide level, but on the more exposed north coast of Cornwall, open to Atlantic storm waves, the relict slope has been largely, and on some sectors (e.g. Watergate Bay, N of Newquay) completely destroyed by Recent marine erosion. The inferred sequence of cliff forms is shown in Fig. 15. Similar features are found in Brittany and S Ireland, on parts of the coast of Washington and Oregon, and on the shores of islands in the Southern Ocean, notably on Auckland Island (Fleming, 1965).

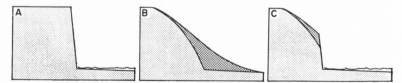

15 *Evolution of slope-over-wall cliffs on coasts subjected to Pleistocene perigla-ciation. A, pre-periglacial sea cliff; B, cliff degraded under periglacial conditions and mantled by rubble drift (shaded black) during Pleistocene glacial phases of low sea level; C, 'slope-over-wall' form produced by marine erosion of the base of the cliff after the sea rose to its present level in Recent times.*

On rocky sectors of the Antarctic coast the formation of degraded slopes by periglacial activity still continues.

The influence of subaerial denudation on the form of cliffed coasts is also important in humid tropical regions, where many coastal rock formations, weakened as the result of decomposition by chemical weathering, do not form steep cliffs. Yampi Sound in N Australia is bordered by low crumbling cliffs of deeply-weathered metamorphic rocks, from which protrude bolder promontories of quartzite, a type of rock less readily modified by chemical weathering (Edwards, 1958). Marine erosion therefore works upon coastal rock formations, the resistance of which is more a function of their response to weathering under humid tropical conditions than of their original lithology. Tricart (1962) has described the persistence of a dolerite headland at Mamba Point in Liberia, where adjacent outcrops of thoroughly weathered granite and gneiss do not form bold cliffs, and the rarity of cliffed coasts in the humid tropics is also apparent in E Brazil (Tricart, 1959).

Cliff forms are also influenced by the geomorphology of the immediate hinterland, in particular the topography which is intersected as the cliff recedes. Other things being equal, cliffs that intersect high ground recede more slowly than cliffs cut into low-lying topography, so that interfluves tend to become promontories between valley-mouth embayments. This has happened on the coast near San Remo, Victoria, where a crenulate coastal outline has developed on the margins of hilly dissected country on Jurassic formations, and rather similar terrain has been sculptured in the same way on the Yorkshire coast N of Flamborough Head. Streams which drain valleys truncated by cliff recession pour out as waterfalls cascading on to the shore, as at Ecclesbourne Glen, on the Sussex coast near Hastings, and deep coastal ravines known as chines in the Isle of Wight and Bournemouth Bay have been cut by runoff from land adjacent to rapidly-receding cliffs. Dry valleys truncated by recession of chalk cliffs on the Sussex coast produce the undulating crest line of the Seven Sisters (Plate 9); locally, valleys that ran parallel to the coast have been dismembered by cliff recession, as in the vicinity of Beachy Head. Recession of cliffs is often measured in linear terms, but it is more useful to take account of variations in cliff height. Williams (1956) recorded 12 m of recession on cliffs 12 m high and 27 m of recession on cliffs 3 m high on a mile-long sector of the Suffolk coast at

9   Chalk cliffs at Seven Sisters on the Sussex coast, England. The cliffs truncate
ridges and valleys on the coastal margin of the South Downs, and there is
an inter-tidal shore platform at their base. In the foreground the Cuckmere
River has built a small delta of sand and gravel, exposed at low tide. (C. T.
Bird)

10   The ruined village at Hallsands on the S Devon coast, England. The village
was built on a coastal ledge (an emerged shore platform) 4–5 m above mean
sea level. A broad shingle beach formerly stood in front of it, but removal of
shingle resulted in accentuated erosion at the cliff base. (C. T. Bird)

Covehithe during the 1953 storm surge, when the loss of material over a 24-hour period amounted to about 300,000 metric tons.

Accumulation of large quantities of beach material on the backshore serves to protect the base of a cliff from wave attack, storm wave energy being expended upon the beach. Smaller quantities of beach material that can be mobilised and hurled on the cliff base during storms accentuate abrasion by waves. Herein lies the risk of removing beach material from the shore. Removal of shingle from the foreshore at Hallsands, in S Devon, for construction work at Plymouth during the eighteen-nineties, has led to accentuated cliff erosion and the destruction of the fishing village of Hallsands, which used to stand on a coastal terrace near the base of the cliffs (Robinson, 1961) (Plate 10). Complete absence of rock material in front of cliffs would leave waves unarmed with abrasive material and capable only of hydraulic action at the cliff base, but as a rule some beach sediment is present, and one way of halting recession of cliffs on weak outcrops is to build and maintain a broad beach by constructing groynes to intercept longshore drifting, augmented if necessary by the dumping of supplementary beach material on the foreshore or updrift. Groynes which intercept longshore drifting may starve the coast downdrift

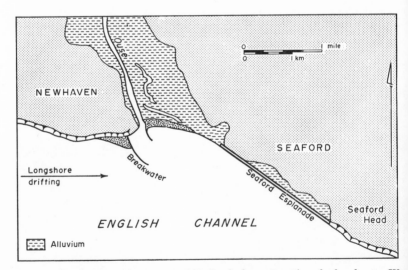

16 *At Newhaven, on the S coast of England, the construction of a breakwater W of the harbour interrupted the eastward shingle drift, so that only a narrow fringe of shingle persists from Seaford Esplanade eastwards*

of sediment, accentuating erosion there. This has happened on the Sussex coast since the building of Newhaven breakwater intercepted eastward shingle drift, leading to depletion of beaches at Seaford, where the seafront is repeatedly damaged in storms, and under Seaford Head where the chalk cliffs are now receding rapidly (Fig. 16). Sea walls may stabilise a shore, but unless a protective beach is maintained in front of them they will need frequent repair.

The outline in plan of a cliffed coast often becomes simplified and smoothed with the passage of time, except where there are marked contrasts in the structure and lithology of coastal rock formations, which may perpetuate irregular outlines as the cliff recedes. Where the coastal outcrops are comparatively uniform, a receding cliffed coast tends to develop an outline in plan related to the prevailing wave patterns, although it is possible for crenulations to develop and persist even on receding cliffs of generally homogeneous material, as on sectors of the Nullarbor coast in S Australia (Plate 11). Where the predominant wave patterns are refracted by offshore topography or adjacent headlands, the cliffed coasts develop gently-curved outlines in plan, much like those on depositional coasts (Chapter V); indeed a curved cliffed sector may pass smoothly into a curved depositional sector, as in some of the asymmetrical embayments on the Victorian coast, notably Waratah Bay W of Wilsons Promontory. Similar curved outlines have developed on the Tertiary cliffs of Bournemouth Bay, in S England, in relation to refracted waves approaching from the SW.

Shore platforms

On the simplest form of cliffed coast, the cliffs are bordered by platforms extending across the shore zone and sloping gently, but not always uniformly, to pass beneath the sea. These platforms are evidently developed and widened as the cliffs recede, and shaped by the action of waves and other marine processes. They extend from high tide mark, at the base of the receding cliff, to a level below and beyond low tide mark, in the nearshore zone, and it is convenient, though not strictly accurate, to describe them as inter-tidal shore platforms (Fig. 13A).

Such platforms are best developed where the coastal rock formations are homogeneous, without structural or lithological variations, but it is difficult to find ideal examples. Vertical cliffs and sloping inter-tidal platforms are found on the Chalk coasts of

11 *Cliffs near the Head of the Bight, in South Australia. The coast consists of almost horizontal Eocene and Miocene limestones. (Australian Tourist Commission)*

12 *Serrated foreshore topography on Palaeozoic rocks at Barraga Point, New South Wales, showing shore platform development along corridors of less resistant rock (O. F. Dent)*

S England and N France (Prêcheur, 1960), but these are not simply the product of wave abrasion (see below, p. 65). The ideal form is sometimes found where structureless sandstone or shale formations have been eroded by wave action, but its development requires a delicate balance between rock resistance and the intensity of wave attack. If the rocks are too resistant, the cliff and the inter-tidal shore platform will not have developed in the time available since the sea reached its present level; if they are too weak they will be unable to sustain a steep cliff profile, and will either show recurrent landslides or will recede until they are reached by the sea only briefly at the highest tides and develop subaerially degraded profiles. Relatively resistant formations may be eroded into steep cliffs and inter-tidal shore platforms on a high wave energy coast, and relatively weak formations may develop these features on a low wave energy coast; under such a balance of conditions, a dynamic equilibrium may be attained, the coastal morphology persisting with parallel recession of cliffs and inter-tidal platforms. In practice there are usually several complicating factors, but the balance of these may sometimes yield a deceptively simple cliff-and-platform profile, as on certain Chalk coasts. More often, variations in structure and lithology in the shore zone persist in irregularities of profile with tracts of platform locally developed, between ridges of harder rock and channels where less resistant outcrops have been excavated (Plate 12).

Waves armed with rock fragments (sand, shingle, cobbles) are undoubtedly powerful agents of abrasion. Without such fragments, waves are capable of only limited abrasion, mainly on soft clay and shale formations. Rock fragments may be of local derivation, eroded from the cliff or the shore platform, or they may have been brought in by longshore drifting from adjacent sectors of the coast, or shoreward drifting from the sea floor. Evidence of wave abrasion can be seen on shore platforms where the more resistant elements of rock persist as reefs and stacks and intervening areas have been scoured out as pools and clefts. Debris used by the waves as an abrasive tool is found littered on the shore platform, particularly after stormy periods; the rock fragments are generally smooth and round as the result of abrasion. Smoothed and scoured abrasion ramps slope gently upward to the base of the receding cliff, which has sometimes been undercut to form a definite abrasion notch. Rock fragments that become trapped in a crevice on a shore platform may be repeatedly moved by wave action in such a way as to excavate circular potholes, smoothly-worn basins containing smoothed and rounded pebbles—clear evidence of the potency of waves armed with abrasive debris.

The depth to which abrasion is active is an important factor in the evolution of shore topography. Sand grains in sea floor sediments off the Californian coast have been shown to become well worn and rounded in depths of less than 10 m below low tide level; in deeper water they show less evidence of wear, and often remain angular. As the wearing of these particles implies the abrasion of the rock surface upon which they rest, this evidence suggests that on coasts exposed to ocean swell, abrasion of the platforms is effective only to a depth of 10 m (Bradley, 1958). On a tideless coast, wave-cut platforms can develop from the water's edge seawards to a depth of 10 m, and with an average inclination of 1° such a platform would be about one-third of a mile (0·5 km) wide. The possible width increases with tide range, and would be about half a mile (0·8 km) with an average slope of 1° where the tide range is 5 m. Where very broad shore platforms are found, they can only be explained as wave-cut platforms if they have developed during a phase of slow marine transgression, the land margin being planed off by waves as the sea rose.

The Chalk cliffs and inter-tidal shore platforms on the English and French coasts bordering the English Channel are to some extent the product of wave abrasion, the waves being armed with flint nodules eroded out of the Chalk (Prêcheur, 1960). Chalk is a relatively homogeneous limestone formation. The best development of vertical cliffs and sloping inter-tidal shore platforms is on the Kent coast near Dover and the Sussex coast E of Brighton, where stratified chalk dips gently seaward. Fresh white chalk exposed on the shore platform and at the base of the cliff after stormy periods is evidence of the wearing and scouring of chalk by waves armed with flints, but other processes are also at work. Cliff recession is marked by recurrent rock falls, particularly after wet weather, or when a spring thaw releases rock masses that had been loosened by frost action along vertical joints and horizontal bedding-planes during the preceding winter. Fans of chalk and flint talus thus produced are subsequently dispersed or consumed by the sea. Examination of boulders that have fallen in this way shows that the chalk surface is pitted by solution, and that the rock has been modified by the physical and biochemical effects of shore flora (chiefly marine algae) and fauna (limpets, mussels, and winkles). These processes play an important part in the reduction and eventual disappearance of the chalk, releasing unworn flint nodules which can then be used in abrasion. Wave action contributes to the wearing and reduction of rock debris and also clears away material that has fallen from the cliffs, thereby ensuring continued recession at the cliff base (Plate 13).

Structural and lithological variations in coastal rock formations can strongly influence the development of cliff and shore platform profiles. In the Sydney district, on the coast of New South Wales, stepped profiles have developed on the outcrops of Triassic sandstones, the steps being 'structural' in the sense that they coincide with the upper surfaces of resistant strata: they are horizontal where the bedding is flat, and inclined where the rocks are dipping (Plate 14). It is possible to follow a particular bench along the shore, down the dip of a resistant sandstone layer, until it passes below sea level, when the bench on the next higher resistant rock outcrop begins to dominate the shore profile (Johnson, 1931; Jutson, 1939). This kind of coastal topography results from storm wave activity, with erosion along bedding-planes and joints and the removal of dislocated rock masses to lay bare a structural bench. Where the rock formations dip gently seaward the profile may resemble an inter-tidal shore platform, passing below low water mark. Where benches or platforms have developed on flat or gently dipping formations above high tide level, it may be tempting to regard them as emerged features, developed during an earlier phase of higher relative sea level, but on high wave energy coasts storm waves may develop 'structural' benches as much as 60 m above present sea level on horizontal or gently-dipping stratified formations (p. 56). A platform above high tide level is more likely to indicate an earlier phase of higher relative sea level if it is cut *across* the local structures of coastal rock formations, and, even then, confirmatory evidence of associated 'raised beach' materials is desirable.

Structural benches of the kind found on the Sydney coast are developed elsewhere on horizontal or gently-dipping sandstones, and occasionally on stratified limestone formations. In Britain, the Old Red Sandstone of northernmost Scotland has benches of this kind, and they are also found locally on the Triassic sandstones of the S Devon coast between Torquay and Sidmouth.

13   *Base of Chalk cliff near Birling Gap, Sussex, England, showing flint shingle and rounded boulders of chalk on the beach and layers of flint in the face of the cliff (C. T. Bird)*

14   *Structural benches on bedded sandstone in the cliffs at Cape Solander, New South Wales (O. F. Dent)*

Much attention has been given in coastal geomorphological literature to horizontal, or nearly horizontal, shore platforms found on many coasts, which truncate local geological structures and cannot be explained in terms of lithological control. Originally observed and studied in New Zealand, Hawaii, and Australia, these are in fact of widespread occurrence on the islands and shores of the Pacific and Indian Oceans, and have also been noted locally on the Atlantic coast. They fall into two main categories: those developed at, or slightly above, mean high tide level ('high tide shore platforms') (Fig. 13B), and those developed slightly above mean low tide level ('low tide shore platforms') (Fig. 13C).

Shore platforms at relatively high levels have been interpreted in various ways (Cotton, 1963). In SE Australia they are typically submerged at high spring tide, and when storms drive waves across them, but on calm days they remain dry at high neap tide (Plate 15). It has been suggested that they are essentially 'storm wave platforms' produced by waves driven across them during storms when the cliff at the rear is cut back; in calmer weather, wave action is limited to the outer edge, which gradually recedes, the width of the platform being a function of the relative rates of front and rear recession. It is difficult to accept this as an explanation for horizontal shore platforms, except in the special case where the platform coincides with the upper surface of a horizontal rock stratum. Attempts to argue that storm waves achieve planation by concentrating their energy at a particular level (Edwards, 1941) are not convincing, since storm waves come in a variety of dimensions, and operate over a height range related to the rise and fall of tides. High tide shore platforms are often as well, or better, developed on sectors of the coast that are sheltered from strong storm wave activity as on the New South Wales coast, where the strongest storm waves arrive from the SE, but the high tide shore platforms are at least as broad and often better developed on the northern sides of headlands and offshore islands. On the more exposed southern sides the platforms show evidence of dissection and destruction at their outer margin and along joints and bedding-planes, and there is evidence that this recession accompanies the development of an inclined platform at a lower level (Bird and Dent, 1966). It appears that wave abrasion, operating alone, tends to develop the simple profile of steep cliff, bordered by 'inter-tidal' shore platform, within restraints imposed by the structure and

15 *High tide shore platforms on the New South Wales coast near Scarborough*

16 *Inner margin of high tide shore platform cut in basalt on Broulee Island, New South Wales, showing the degraded bluff*

lithology of coastal outcrops. Horizontal platforms truncating local geological structures cannot be explained in terms of storm wave attack, which appears to have a secondary and modifying influence on these features.

It has been suggested that shore platforms at or slightly above mean high tide level are the product of wave abrasion at an earlier phase when sea level was higher (Fairbridge, 1961), the platforms merely being kept fresh by the surf that washes over them at high spring tides and during storms. The hypothesis accords with the idea that the Holocene marine transgression attained a maximum a few feet above present sea level before dropping back during a phase of 'Recent emergence', and, where high tide shore platforms are backed by degraded cliffs which have not been kept fresh by marine attack, the evidence for this view appears strong (Plate 16); dissection of the outer edge of the platforms can then be interpreted as the result of the cutting of a new platform by wave abrasion at a lower level following emergence. However, it is difficult to explain why high tide shore platforms are horizontal, or nearly so, whereas the platform being cut by wave abrasion at a lower level has a seaward inclination. Moreover, the cliffs behind high tide platforms are not always degraded. Some are clearly active and receding, but the degradation of others could be simply a function of intermittency of storm wave attack, and a relatively long phase of quiescence since they were last trimmed back by the sea. It is necessary to consider what other processes are at work on these cliffs and platforms before deciding if it is necessary to invoke an episode of higher sea level in Recent times to account for their development.

Examination of shore features in the vicinity of high tide shore platforms yields evidence that weathering processes are at work on coastal rock outcrops unprotected by a soil or vegetation cover in the zone at and above mean high tide level. Superficial decomposition of exposed rock surfaces results from repeated wetting and drying, accompanied by salt crystallisation, in the zone subject to the action of spray. Tricart (1959) claimed that salt spray weathering is a dominant process in shore platform development on the tropical coast of Brazil, where absence of coarse detritus impedes abrasion by waves and high insolation rapidly dries off rock surfaces wetted by saline spray. It is, however, difficult to dissociate the physical effects of wetting and drying from the physico-

chemical effects of crystallisation from drying spray. Salt crystallisation probably has specific corrosive effects, and may accelerate
weathering compared with wetting and drying in freshwater
environments: laboratory tests suggest that this is so, but the
problem requires further investigation in the field.

Sandstone outcrops subject to wetting by seaspray and rainfall,
and subsequent drying, become pitted and honeycombed as sand
grains are loosened by the decomposition of the cementing material
that formerly bound them. Other fine-grained rock formations,
such as siltstones, mudstones, shales, schists, phyllites, and basalts,
are subject to similar superficial decomposition and weathering
effects. This kind of weathering is not effective at lower levels,
where rock formations are permanently saturated by sea water,
and so wave action washes away the disintegrating material above a
certain level, gradually laying bare a platform which coincides with
the upper level of permanent saturation. On the coast of SE
Australia it is the fine-grained rock outcrops, which show evidence
of pitting and cavity formation indicating that the exposed rock is
gradually being disintegrated by weathering processes, that have
shore platforms developed at or slightly above mean high tide level.
Pools and channels on the platform surface become enlarged and
integrated as their overhanging rims recede, and gradually the rock
surface is 'peeled off' down to a level which remains intact because
it is permanently saturated (Plate 16). The process has been
described as 'water-layer weathering', and as the level of saturation
need not coincide with bedding-planes it offers a mechanism by
which coastal planation may operate, forming platforms which
may transgress local geological structures (Hills, 1949). It also
accounts for the fact that high tide shore platforms are almost
horizontal, often with a raised rim at the outer edge which is
permanently saturated by breaking waves. Finally, this type of
weathering is less effective on massive coastal outcrops of granite
and quartzite on the SE coast of Australia, which explains why
some headlands, such as the granite mass of Wilsons Promontory
in Victoria, are not bordered by high tide shore platforms (Plate 17).

Water-layer weathering explains many features of high tide
shore platforms, but it cannot explain them entirely. Occasional
storm wave activity is necessary to sweep away the disintegrated
rock material, and to attack and rejuvenate the base of the cliff at
the rear of these platforms. On extremely sheltered sectors of the

17 *Plunging cliffs of granite at the southern end of Wilsons Promontory, Victoria (Australian News and Information Bureau)*

18 *Part of the low tide shore platform cut in Pleistocene calcarenite on the Victorian coast near Portsea, showing the notch and overhanging visor at the base of stacks*

coast it is possible that the weathered material would not be carried away sufficiently for the high tide shore platform to develop, and on sectors exposed to very frequent storms it is possible that abrasion would be sufficiently rapid to destroy the high tide platform as quickly as the removal of weathered debris laid it bare. The width of such a platform is determined by the relative rates of recession of the cliff at the rear (by removal of weathered material and occasional storm wave abrasion) and the frontal margin (by more continuous wave action on permanently saturated and unweathered rock). Where nearshore rock debris is available, accelerated dissection of the frontal margin may lead to the development of an inter-tidal shore platform. On the New South Wales coast high tide shore platforms give place to inter-tidal platforms on exposed sectors where wave abrasion is facilitated by the presence of locally-derived shingle (Bird and Dent, 1966). The nature and extent of shore platforms on such a coast vary in relation to a number of factors: lithology, structure, weathering processes, nearshore topography, wave régime, tidal range, and the availability of abrasive debris—factors that vary intricately on a coast of irregular configuration, with local variations in aspect. As the sea stood at higher levels relative to the land at certain phases during Quaternary times, existing shore platforms have evidently developed as the result of down-wasting and planation of presumably similar features that developed earlier at higher levels. Locally, fragments of older and higher platforms may persist as emerged terraces. High tide shore platforms on the SE coast of Australia are related to present sea level and wave conditions and it is not necessary to invoke an episode of higher sea level to account for them; on the other hand they cannot be taken as evidence that a higher stillstand did not occur in Recent times.

The development of shore platforms by the washing away of disintegrated material where rock formations have been weathered down to a certain level was described by Bartrum who termed these platforms of 'Old Hat' type, with reference to an island off the NE coast of New Zealand (Cotton, 1963). In N Australia, Edwards (1958) similarly explained broad platforms on the shores of Yampi Sound in terms of removal by wave action of rock that had been deeply weathered, under tropical humid conditions, to expose a shore platform of unweathered rock. Climatic factors may also have influenced the development of strandflats, the extensive shallow-water and partly emerged platforms of problematical origin found on the fiord coasts of Norway, Spitzbergen, Iceland, and Greenland. They are up to 64 km wide locally, and numerous rocky islands rise above them. It is possible that they developed as the result of plucking and disintegration of coastal rock outcrops by shelf ice, followed by the sweeping away of debris by wave action when the ice melted. Alternatively, they may be the outcome of prolonged coastal periglaciation. The Norwegian strandflat appears to be a relict feature (Tietze, 1962), but the processes which produced it have not yet been identified on actively glaciated or periglaciated Arctic or Antarctic rocky coasts.

Low tide shore platforms may be defined as horizontal, or almost horizontal platforms exposed only for a relatively brief period when the sea falls below mean mid-tide level. They are best developed on certain limestone coasts, where they are broad and almost flat, except for an inclined ramp towards the rear, leading up to the cliff base, which frequently has a notch overhung by a visor (Plate 18). There is sometimes a slightly higher rim at the outer edge formed by an encrustation of algae (chiefly *Lithothamnion* species) in a zone that is kept wet by wave splashing even at low tide (Fig. 17).

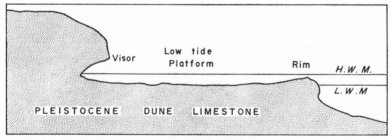

17   *Low tide shore platform, as developed on Pleistocene dune limestone (aeolian calcarenite) on the S and W coasts of Australia*

As low tide shore platforms of this kind are typically developed on limestone formations, it is evident that solution of limestone (chiefly calcium carbonate) by water, in the presence of carbon dioxide, is an important factor in their formation (Wentworth, 1939). The dissolving of limestone is represented by the equation:

$$CaCO_3 + H_2O + CO_2 \rightarrow Ca(HCO_3)_2$$

the rock passing into solution as calcium bicarbonate, $Ca(HCO_3)_2$. Solution of limestone can also be accomplished by rainwater and by ground-water seepage provided that the water contains dissolved carbon dioxide and is not already saturated with calcium bicarbonate in solution. In fact, sea water off limestone coasts and ground water emerging from limestone cliffs is normally saturated with dissolved calcium bicarbonate, and thus incapable of dissolving more limestone on the shore. Rain water, on the other hand, is rich in dissolved atmospheric carbon dioxide, and is capable of dissolving away a limestone surface. This is a contributory factor in the development of limestone shores exposed to rainfall at low tide, but the fact that low tide platforms are at least as well developed on the coasts of arid regions indicates that some other factor must be dominant.

The problem of how sea water, already saturated with calcium bicarbonate, can achieve further solution of limestone in the shore zone was analysed by Revelle and Emery (1957) on Bikini Atoll. They found that there were marked diurnal variations in the dissolving capacity of sea water, related in part to variations in the temperature and carbon dioxide content of water in the shore zone. Nocturnal cooling of sea water increases its capacity to take up carbon dioxide (which is more soluble in water at lower temperatures), permitting it to dissolve more limestone, and the biochemical activities of marine organisms lead to the production of carbon dioxide (from plant and animal respiration) which is used by the plants (chiefly algae) in photosynthesis by day but which accumulates at night when the absence of sunlight halts photosynthesis. The nocturnal increase in the acidity of coastal water permits limestone to be taken into solution. Additional calcium carbonate dissolved during the night will be precipitated by day, when temperature rises and photosynthesis revives, but the precipitated sediment is likely to be carried away from the limestone shore by waves and currents. As the nocturnal solution processes operate mainly within the tidal zone, they will remove calcium carbonate in solution down to a level at which permanently saturated and submerged limestone is evidently dissolved more slowly, if at all, by sea water, so that a low tide platform tends to develop. According to Hills (1949), shore platforms developed on calcarenites coincide with a horizon of secondary induration, where cementing calcite has been precipitated from percolating ground water encountering carbonate-saturated sea water: the calcarenite is less resistant both above and below this horizon, which thus becomes a structural factor in shore platform evolution.

As with the water-layer weathering process effective on the coastal rock formations where high tide shore platforms have developed, solution processes are accompanied by occasional storm wave activity, which sweeps away precipitated calcium carbonate, and, if armed with suitable rock fragments, achieves abrasion of the outer margin of the platform and of the ramp and cliff at the rear. Chalk coasts, discussed previously, would evidently have developed similar low tide shore platforms, were it not for the availability of associated flints, which have been used in wave abrasion, leading to the development of a broad inter-tidal platform. The notch and visor feature typical of many limestone coasts is evidently not simply an abrasion notch; it is as well developed on sheltered sectors of the coast as on sectors exposed to strong wave action, and extends uniformly around the stacks in the foreground of Plate 18. Exposure to strong storm wave activity is evidently inimical to the development of this feature, for storm waves have been observed to snap off visors that had developed under quieter conditions. Hodgkin (1964) found that maximum solution of coastal calcarenites (about 1 mm/year) takes place just above mean sea level, where the notch develops, indicating that it is essentially the outcome of solution processes. Marine organisms also contribute to the etching away of limestone in the shore zone, and Emery (1960) has indicated that in favourable conditions biochemical processes can consume rock at least as quickly as physical and purely chemical erosion. Some would go as far as to suggest that the notch has been almost literally eaten out by the shore fauna that occupies this horizon.

Shore morphology on limestone coasts shows variations related to zonal climatic factors. Guilcher (1953) has outlined the characteristic features of limestone (excluding Chalk) coasts in the cool temperate, warm temperate, and tropical zones. In cool temperate regions (S Ireland, S Wales) limestones show pitting and honeycombing in the upper (spray) zone, shallow flat-floored pools with overhanging rims, tending to widen and coalesce as a platform, in the shore zone, and a more irregular network of ridges and pinnacles (coastal lapies) in the lower part of the shore zone (Fig. 18A). These predominantly solution forms are obliterated where waves

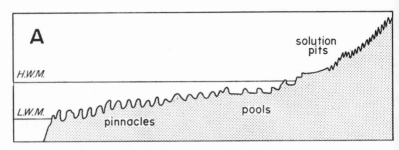

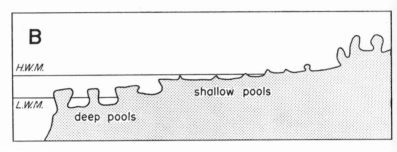

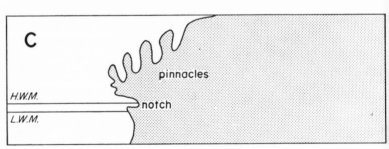

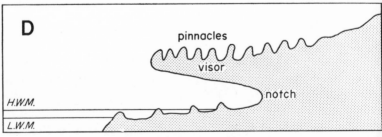

18   *Typical zonation of corrosion forms on coastal limestone (after Guilcher, 1958):
A, cool temperate regions (British Isles); B, warm temperate regions,
microtidal (Mediterranean); C, warm temperate regions, mesotidal (Morocco); D, tropical regions (Pacific atolls)*

armed with rock debris abrade the shore zone, and lithological variations introduce further complications, as on the stratified Jurassic limestones of the Dorset coast. In warm temperate regions Guilcher recognised two kinds of zonation, both with coastal lapies in the upper (spray) zone (Plate 19). On mesotidal coasts (Morocco, Portugal) the shore zone consists of flat-floored pools becoming larger and deeper towards low tide mark; on microtidal coasts (Mediterranean) there is a notch, with an overhanging visor, and a narrow shore ledge exposed at low tide (Fig. 18 B and C). The zonation on Australian calcarenite shores in a warm temperate environment is similar except that the shore platform is somewhat broader. Guilcher (1958b) found that the warm temperate zonation gave place to the cold temperate zonation from S to N around the Bay of Biscay. In tropical regions, especially on emerged coral, the pinnacles are developed on top of an overhanging visor, and the flat-bottomed pools occupy a ledge exposed at low tide (Fig. 18D).

There are similar variations in the characteristic forms of cliffed coast on granites and basalts, related to the weathering régime under different climatic conditions, but the details of these have not yet been worked out.

Plunging cliffs

Plunging cliffs (Fig. 13D) are cliffs that pass steeply beneath low tide level without any development of shore platforms. These have several possible explanations. Plunging cliffs can be produced by Recent faulting, the cliff being the exposed plane of the fault on the up-throw side, the down-thrown block having subsided beneath the sea. The cliffs along the line of the Wellington fault, on the W shore of Port Nicholson in New Zealand, have been explained in this way (Cotton, 1952); they slope at about 55°, show little evidence of marine modification at the shore level, and descend to the 12 fathom (about 22 m) submarine contour, close inshore and parallel to the coastline. Tectonic subsidence of coasts may lead to the development of plunging cliffs, possibly with former shore platforms submerged beneath low tide level; Cotton (1951b) suggested that the plunging cliffs of Lyttleton Harbour and Banks peninsula (Fig. 45, p. 123) originated as the result of continuing subsidence of this area during and since the Holocene marine transgression. Coasts built up recently by volcanic activity,

19   *Coastal lapies on a limestone shore at Two Mile Bay, Victoria*

as on the island of Hawaii, show plunging cliffs on sectors where there has not yet been time for shore platforms to develop. Absence of shore platforms on very sheltered sectors of coast bordering rias and fiords, and on sectors where the coastal rock outcrops are extremely resistant, may simply be due to the fact that the period of up to 6000 years since the sea attained its present level has been too brief for marine processes to develop platforms. In Australia the plunging granite cliffs of Wilsons Promontory (Plate 17) are evidently too resistant for shore platforms to have developed within this relatively short period; the cliffs are the partly-submerged slopes of granite hills and mountains. On the Nullarbor coast, there are sectors where the vertical cliffs of homogeneous Eocene and Miocene limestones pass beneath low tide level without shore platforms, possibly because the great vigour of storm wave activity on this coast is achieving abrasion down to a level below low tide level as these cliffs recede (Plate 11).

The morphogenic system leading to the evolution of cliffed coasts can be analysed in terms of the effects of geological structure and lithology, the action of marine and subaerial processes, tidal conditions, and the inheritance of changing levels of land and sea and changing climatic conditions. The tempo of change on a cliffed coast shows marked variations from place to place, even on adjacent flanks of an embayment or headland, according to rock resistance and degree of exposure to marine attack. At one extreme there are sections of coast, like Wilsons Promontory, which have changed very little in the period since the Holocene marine transgression brought the sea to its present general level. At the other extreme there are sections of rapid cliff recession, as on the coast of W Victoria, near Port Campbell, and sectors of the E and S coasts of England, where relatively weak Tertiary formations confront stormy seas, and changes can be measured by referring to old maps and photographs (Fig. 19). Between these extremes there are many sections of cliffed coast that are developing slowly and showing variations which can be analysed to show the respective roles of rock resistance, wave attack, and coastal weathering processes in their developing morphology.

Studies of the evolution of cliffed coasts have generally been largely qualitative, concerned with identifying and attempting to evaluate the various factors and processes that have been at work. In quantitative terms it would be possible, given adequate data on

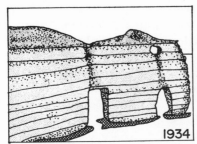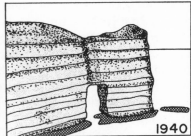

19   *The betrunking of Elephant Rock, near Port Campbell, Victoria (after Baker, 1958)*

wave direction and height in relation to tidal levels, to derive a time-integrated expression of the wave energy profile on a cliffed coast. Profiles of this kind have been used in engineering studies, notably in the design of sea walls, but there are difficulties in attempting to use such data to explain the configuration of a cliffed coast. Existing configuration is one of the important factors determining the wave energy profile so that there is a danger of circular argument. Moreover, as has been indicated, erosion of cliffed coasts is the outcome of a variety of processes, including weathering, solution, and biochemical activity, which are largely independent of wave energy profiles. It is perhaps because of these difficulties that satisfactory quantitative appraisals of the evolution of cliffed coasts have yet to be made.

V

BEACHES, SPITS, AND BARRIERS

Beaches are accumulations of sediment deposited by waves and currents in the shore zone. In terms of the Wentworth scale of particle diameters (Table 1) they are typically composed of sand

TABLE 1

Wentworth size class	Particle diameter	φ scale
Boulders	> 256 mm	below −8φ
Cobbles	64 mm — 256 mm	−6φ to −8φ
Pebbles	4 mm — 64 mm	−2φ to −6φ
Granules	2 mm — 4 mm	−1φ to −2φ
Very coarse sand	1 mm — 2 mm	0φ to −1φ
Coarse sand	$\frac{1}{2}$ mm — 1 mm	1φ to 0φ
Medium sand	$\frac{1}{4}$ mm — $\frac{1}{2}$ mm	2φ to 1φ
Fine sand	$\frac{1}{8}$ mm — $\frac{1}{4}$ mm	3φ to 2φ
Very fine sand	$\frac{1}{16}$ mm — $\frac{1}{8}$ mm	4φ to 3φ
Silt	$\frac{1}{256}$ mm — $\frac{1}{16}$ mm	8φ to 4φ
Clay	< $\frac{1}{256}$ mm	above 8φ

Note: The φ (phi) scale is a logarithmic scale of particle diameters defined by Krumbein as the negative logarithm to base 2 of the particle diameter in millimetres: $\varphi = -\log_2 d$

or shingle. The proportions of each grain size can be determined by mechanical analysis, when a known weight of dried beach sediment is passed through a succession of sieves of diminishing mesh diameter, and divided into size grades which are weighed separately. As a rule, samples of beach sediment are well sorted, in the sense that the bulk of a sample falls within a particular size grade, as indicated by the graph in Fig. 20. Statistical parameters based on mechanical analysis indicate that the grain size distribution of beach sediment is commonly asymmetrical, and negatively-skewed, the mean grain size being coarser than the median; evidently the winnowing effect of wave action reduces the relative proportion of fine particles (Mason and Folk, 1958). Such parameters may assist in the identification of former beach deposits, permitting a distinction from positively-skewed grain size distributions typical of fluvial, aeolian, and lagoonal deposits (Friedman, 1961). Chappell (1967) has used this method to identify shoreline sediments in the Quaternary deposits of the W coast of North Island, New Zealand.

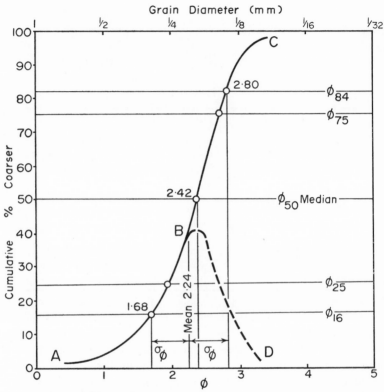

20   *Graphic representation of a grain-size analysis of a beach sand. Measured in*
      φ *units, the median diameter, Md_φ (50th percentile) is 2·42, the 16th per-*
      centile (φ_{16}) 1·68, and the 84th percentile (φ_{84}) 2·80. Values for the mean
      (M_φ), sorting (σ_φ), and skewness (α_φ) are calculated as follows:

$$Mean = M_\varphi = \tfrac{1}{2}(\varphi_{16} + \varphi_{84}) = \tfrac{1}{2}(1·68 + 2·80) = 2·24$$
$$Sorting = \sigma_\varphi = \tfrac{1}{2}(\varphi_{84} - \varphi_{16}) = \tfrac{1}{2}(2·80 - 1·68) = 0·56$$
$$Skewness = \alpha_\varphi = \frac{M_\varphi - Md_\varphi}{\sigma_\varphi} = \frac{(2·24 - 2·42)}{0.56} = -0·32$$

(i.e. the mean (2·24) is coarser *than the median (2·42) and the grain-size distribution is negatively-skewed).*

Beach sediments composed of larger particles tend to develop steeper gradients (Table 2), chiefly because of their greater permeability. Wave swash sweeps sediment forward on to a beach, and backwash tends to carry it back, but the greater permeability of shingle and coarse sand beaches diminishes the effects of back-wash, leaving swash-piled sediment at relatively steep gradients. Fine sand beaches are more affected by backwash, and have gentler slopes, often of firmly-packed sand across which it may be possible to drive a car.

TABLE 2

Wentworth size class	Mean slope of beach face
Cobbles	24°
Pebbles	17°
Granules	11°
Very coarse sand	9°
Coarse sand	7°
Medium sand	5°
Fine sand	3°
Very fine sand	1°

Source: Shepard (1953)

The origin of beach sediments

The nature of beach sediment is clearly related to the nature of material derived from the adjacent cliffs and foreshore, or brought in from offshore or alongshore sources. Shingle beaches are found where coastal rock formations yield debris of suitable size, such as fragments broken from thin resistant layers in stratified sedimentary rocks, or eroded out of conglomerates or gravel deposits, or derived from intricately-fissured igneous outcrops. Much of the shingle on the beaches of SE England has been derived, directly or indirectly, from flint nodules eroded out of the Chalk; some of it has had a long and complex Tertiary and Quaternary history. On the world scale, coasts bordering regions which are, or have been, subject to glaciation and periglaciation have relatively abundant supplies of rock fragments derived from the erosion of rubble drift, gravelly moraines, or drumlins, for the formation of shingle beaches; elsewhere sandy beaches are predominant unless a shingle source exists locally. Shingle is rare on tropical coasts, except where it is derived from coral (Chapter IX), or where torrential rivers deliver gravels to the shore, as at Lae, in New

Guinea, which has a cobble beach supplied with gravels from the Markham River.

Sandy beaches may be supplied with sand eroded from coastal arenaceous outcrops, as in Bournemouth Bay on the S coast of England, and on the eastern shores of Port Phillip Bay, Australia, both of which have beaches derived mainly from receding cliffs of Tertiary sandstone. At Bournemouth the progressive extension since 1900 of a promenade on the seafront has halted cliff recession and cut off the sand supply; the beach is gradually diminishing on a wasting foreshore, with little if any replenishment of sand carried away by longshore drifting.

Sand may also be delivered to the shore by rivers, as on the coast of S California, where the beaches consist of sand of fluvial origin that drifts southwards from river mouths, accumulating against the northern flanks of promontories (such as Point Dume), or lost into the heads of the submarine canyons that run out from the southern ends of several embayments. There are thus distinct compartments of beach material on this coast, and when the fluvial sand supply is intercepted, following construction of reservoirs in the river valleys, beaches are reduced and shoreline erosion accelerated south of the river mouths (Emery, 1960). Fluvial catchment yields, augmented as the result of man-induced erosion (following deforestation, overgrazing, or unwise cultivation), can add to the supply of sand delivered to beaches such as these.

The large-scale sandy beaches on the SE coast of Australia cannot be explained in these ways. The Ninety Mile Beach in Victoria, for example, is not adjacent to eroding sandstone cliffs; it forms part of a barrier system that seals off river mouths as coastal lagoons in which the fluvial sand supply is intercepted. It appears that the beach sand here has been swept shoreward from the sea floor, during and possibly since the Holocene marine transgression. The possibility that sand is still being delivered from the sea floor receives support from experimental work in wave tanks, which show that waves moving towards a coast generate a shoreward current at depth, close to the sea floor (Bagnold, 1947). In addition, King (1953) has shown that winds blowing offshore set up a shoreward movement of water close to the sea floor. Both processes are capable of delivering sediment to a beach, but the depth from which material can be carried shoreward varies in relation to sea floor topography, wave conditions, and the size, shape and specific gravity of the available sediment. Submarine vegetation is also a factor, sediment flow being inhibited where such plants as *Zostera*, *Posidonia*, and *Cymodocea* form luxuriant growths, as in parts of the Mediterranean and off the South Australian coast. In favourable conditions it appears that sea-floor sediment can be drifted shoreward from depths of up to 18 m. In the Mediterranean, Van Straaten (1959) found that waves are moving sand on to beaches on the French coast from a depth of 9 m, and in areas subject to ocean swell the process will be effective from greater depths than this. Off S California there is only limited and occasional movement of sea-floor sand by ocean swell in depths exceeding 18 m, indicating the probable maximum limit. Sand dumped in water 12 m deep off the New Jersey coast failed to move shorewards to replenish an eroding beach (Harris, 1955), and was therefore beyond the limit for shoreward drifting in that environment. In view of these limitations, it seems probable that the bulk of shoreward sand drifting on the SE Australian coast took place during the Holocene marine transgression, when the near-shore zone migrated shoreward over unconsolidated deposits stranded during the previous sea regression (Bird, 1965).

In general, sandy beaches are supplied partly by material eroded from adjacent parts of the coast, partly by fluvial sediment, and partly by sand carried shoreward from the sea floor, the relative proportions being determined by local conditions. In addition, quantities of sand may be blown from the land into the sea, particularly on desert coasts, and thence delivered to the beach. The dynamics of a beach sediment system are thus determined by the balance of nourishment from these various sources against alongshore and offshore losses, and the removal of sand by wind to build landward dunes. Beach dynamics can be analysed in scale model tanks with simulated waves and currents (Saville, 1950), although there are difficulties over the scaling down of natural sediment calibre to fit model conditions. Sand grains can be used in a model to represent a shingle beach, but the representation of a sandy beach to scale may involve the use of silts, which have different physical properties, and may not reproduce the correct response of sand to wave and current action.

More recently there have been theoretical studies of the dynamics of natural beach systems. Miller and Zeigler (1958) presented a theoretical model relating nearshore processes to sediment characteristics on shores where an equilibrium has been attained between process and form. They considered the dynamics of shoaling waves entering shallow water, of turbulence in the breaker zone, and of sheet flow in the zone of swash and backwash, and postulated trend maps of sediment size (maximum in the breaker zone) and sorting (best on the shoreward side of the breaker zone) related to these dynamics. The theoretical model was tested against maps of median grain size and the sorting of nearshore and foreshore sectors of four American beach systems, with generally good results. Krumbein (1963) proposed a process-response model for the analysis of beach phenomena in which the process side is influenced by the energy factor (waves, tides, currents), the geological material factor, and the overall geometry of a coastal sector, and the response side is the form of the beach and the nature of its component sediments. In quantifying this model, 'feedback' is recognised: the beach morphology and sediments influence the processes at work on them. Numerous variables can be defined and measured, and their relationships investigated with the aid of a computer. Application of this kind of analysis soon reveals the complexity and variability of natural beach systems.

Composition of beach sediments

Beach sands of terrigenous origin generally have quartz as the most abundant mineral, accompanied by felspars, mica, and varying proportions of heavy minerals. Sand supplied from erosion of adjacent cliffed coasts or brought down from river catchments reflects the mineral composition of the source areas. The concentration of rutile, ilmenite, and other heavy minerals in beach sands on the N New South Wales and S Queensland coast has been related to fluvial supply from the Mesozoic sandstones of the Ipswich-Clarence basin, sandstones derived earlier from denudation of Permian granites on the New England plateau, within which these heavy minerals originally formed (Gardner, 1955).

Sand supplied from the sea floor may earlier have been of terrigenous origin, laid down during phases of low sea level and subsequently reworked by wave action during marine transgressions. On the New South Wales coast, sand delivered from the sea floor has smaller proportions of the less resistant felspar and mica grains than freshly supplied fluvial sand, the beaches being predominantly quartzose away from river mouths (Bird, 1967b).

Calcareous sands derived from shell fragments or coral debris are common on oceanic island and tropical coasts and are also indicative of shoreward transportation. In temperate regions they are found where shelly organisms are abundant offshore, as on the Lands End peninsula, the Channel Islands, and in the machair of the Outer Hebrides, where beaches of broken shell sand are the source of coastal tracts of calcareous wind-blown sand. Shelly beaches are also common on the shores of estuaries and lagoons: they are found on the shores of the Firth of Forth near Edinburgh. Beaches bordering desert coasts are often calcareous, probably because the fluvial yield of terrigenous sediment has been meagre: the Eighty Mile Beach, bordering semi-arid NW Australia, exemplifies this.

Calcareous beach sands dominate the S and W coasts of Australia, whereas quartzose beach sands are characteristic in the SE. The contrast has never been satisfactorily explained. The junction between the two provinces occurs in Bass Strait, calcareous sands being dominant W of Wilsons Promontory and on the W coasts of King Island and Flinders Island, and quartzose sands to the E. The calcareous sands are derived partly from earlier calcareous sandstones and partly from shell disintegration; the quartzose sands are fundamentally of granitic origin. Even on the Queensland coast, behind the Great Barrier Reefs, beach sands are mainly quartzose, supplied by such rivers as the Fitzroy and the Burdekin, with relatively small admixtures of calcareous reef debris.

On volcanic islands, such as Hawaii, black sand beaches occur, consisting mainly of basaltic fragments. Volcanic pumice is a common, and often far-travelled, constituent on many Pacific coast beaches. Coasts adjacent to lumbering areas in W Canada and N Russia are piled high with sawn timber washed up on the shore, but the most unusual beach material known to this author is on Brownsea Island, in Poole Harbour (Dorset, England) where wave action has distributed the wastes of a former pipeworks to build a beach of broken subangular earthenware.

There are lateral variations in the composition of beach material, particularly in the vicinity of eroding headlands, where the proportion of locally-derived material may be high, and near river mouths, where a larger proportion of fluvial sediment is likely to be present. Chesil Beach, on the S coast of England, consists mainly of flint pebbles, but at the SE end, adjacent to cliffs of Portland and Purbeck Limestone, cobbles and pebbles of limestone are mixed with the flint shingle. In Encounter Bay, on the South Australian coast, the beach consists almost entirely of calcareous sand towards the SE end near Kingston (almost 90 per cent calcium carbonate), the proportion of quartz sand increasing northeastwards to Goolwa at the mouth of the Murray (calcium carbonate content less than 10 per cent); a lateral variation that may reflect the supply of terrigenous material near the mouth of the Murray, diluting the calcareous sand of marine origin on that sector (Sprigg, 1959).

Wave action achieves wearing and attrition of pebble and sand grains, the pebbles tending to develop a rounded form, often slightly flattened (i.e. thinner at right-angles to the principal plane), the sand grains becoming smooth, well-rounded, and highly polished. Attempts have been made to distinguish between beach and dune sands in terms of the shape and roundness characteristics of the component grains (Shepard and Young, 1961; Moss, 1962, 1963), and recently electron microscopy has been applied to the study of the distinctive surface textures and markings on quartz grains from beach and dune environments (Biederman, 1962). These techniques can supplement statistical analyses of grain-size distributions mentioned previously in determining former shore-lines at the upper limits of beach sands in stratified deposits.

Lateral movements of beach material
Waves that arrive at an angle to the shore produce a transverse swash, running diagonally up the beach, and a backwash that retreats directly seaward. Sand and shingle are edged along the shore by waves that break in this manner ('beach drifting') and at

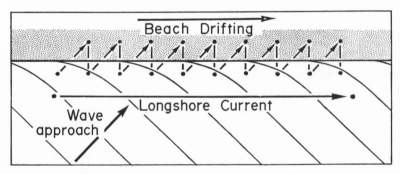

21   *Longshore drifting, the flow of coastal sediment produced by wave and current action when waves approach at an angle to the shoreline*

the same time longshore currents developed by waves in the near-shore zone drive sand along the sea floor (Fig. 21). The combined effect of these processes is to produce 'longshore drifting', the sediment flowing first one way, then the other, according to the direction from which the waves approach. The resultant drift over a period may be negligible, or there may be a definite long-term drift in one direction rather than the other, usually indicated by the longshore growth of spits and the deflection of river mouths (Fig. 39, p. 117).

On the N Norfolk coast a predominant westward drift resulting from the prevalence of waves from the NE is indicated by the westward growth of recurved spits at Blakeney Point (Fig. 30) and Scolt Head Island (Steers, 1960). When waves arrive from the NW an eastward counter-drift is produced, and in some years the net drifting of coastal sediment has been eastward rather than westward. The evidence of spit growth, however, indicates that the long-term resultant has been westward.

On coasts exposed to ocean swell, longshore drifting is produced only if the swell is incompletely refracted so that it arrives at an angle to the beach, or if waves generated by onshore winds move at an angle through the swell to reach the shoreline obliquely. On the N New South Wales and S Queensland coast the predominant SE winds produce waves which set up a northward drift of sand on shores shaped by refracted ocean swell. This has led to the development of major depositional features at Moreton, Strad-broke, and Fraser Islands during Pleistocene and Recent times (see p. 141). On some coasts the pattern of drifting varies season-ally. Boomer Beach in California is a well known example of this, waves from the SW moving sand northward in summer and waves from the NW driving it back to the S in winter. Similar features have been noted on the beaches bordering Port Phillip Bay, where west-facing beaches show the same seasonal response to SW waves in summer and NW waves in winter (Bowler, 1966).

Sand and shingle can 'by-pass' the mouths of rivers and tidal entrances (Bruun and Gerritsen, 1959) and be carried round head-lands by longshore drifting in the nearshore zone (Trask, 1955), but large cliffed promontories that end in deep water are likely to interrupt the longshore flow and act as boundaries between dis-tinct compartments of coastal sediment. Within such compart-ments, beach material moves to and fro along the shore and backwards and forwards at an angle to the shore, but it does not pass the bordering promontories, and there can be striking differences in the nature of beaches in adjacent embayments. Examples of this are found on the coast of Cornwall, where the mineralogical composition of beach sands varies from one cove to another, and on the W coast of Wilsons Promontory (Plate 20), where three adjacent coves (Leonard Bay, Norman Bay, and Oberon Bay) have strongly quartzose, mixed quartzose/calcareous, and strongly calcareous beach sands respectively.

An understanding of patterns of longshore drifting is of great importance in the location, design, and maintenance of harbours. Sediment carried along the shore may be deposited in a harbour entrance, necessitating repeated dredging, or built up alongside breakwaters in such a way as to interrupt longshore sediment flow and lead to reduction of beaches and accelerated erosion on the downdrift shore. This problem is well known on many coasts; it has happened, for example, at Santa Barbara and Santa Monica on the Californian coast, and at the harbours of Durban, Lagos, and Madras. On the S coast of England the building of a breakwater to protect Newhaven harbour intercepted the eastward drift of shingle, depleting the beach and resulting in excessive storm damage and coast erosion at adjacent Seaford (Fig. 16). At Durban, and several American harbours, efforts are made to carry or pipe intercepted sediment past harbour entrances, depositing it to replenish the beach on the downdrift shore. At Port Hueneme, California, an offshore breakwater has been built parallel to the coast on the updrift side of the harbour entrance to create a pattern of wave refraction that concentrates sand accumulation behind the breakwater, where it is excavated at the rate of 500,000 cubic yards (about 375,000 m³) a year by a floating dredge and ferried past the entrance to replenish wasting beaches on the downdrift shore (Johnson, 1957). Where seasonal patterns of longshore drifting occur, as in Port Phillip Bay, breakwaters built to form boat harbours collect the southward drift in winter in a situation where it cannot be returned northward in summer. The harbours thus become filled with sand and the adjacent beaches erode away. This kind of problem indicates the difficulties that arise when coastal constructions interfere with patterns of longshore drifting. Knowledge of sediment source and flow patterns is essential if constructions are to be designed in such a way as to conserve the

20 *Norman Bay, Wilsons Promontory, Victoria, showing refracted ocean swell entering the bay, parallel dunes immediately behind the beach, and older parabolic dunes farther inland*

21 *Ridge and runnel foreshore near Bowen on the Queensland coast. The high island in the background is Gloucester Island*

balance of erosion and deposition on adjacent shores (Beach Erosion Board, 1961).

Tracing sediment flow

In recent years progress has been made with methods of following the movement of sand on beaches and across the sea floor by means of tracers. A quantity of tracer material is introduced at one point, and a search is made to see where it goes. In order that it shall move in the same way as the natural sediment, the tracer particles must have a similar range of size and shape, similar hardness, and the same specific gravity as the sediment already present; they must also be identifiable after they have moved along a beach or across the sea floor. Some tracer becomes buried, and hidden from view within a beach, and a proportion may be carried offshore.

Naturally-coloured and artificially-coloured sand have been used in simple tracer experiments, but there are difficulties in observing coloured grains when they form only a small proportion of the sand on a beach. These have been partly overcome by the use of sand 'labelled' with a coating of colloidal substances containing one of several commercially available fluorescent dyes (Yasso, 1966). Sand labelled in this way is dumped on the beach, and its movements can be traced by locating the dyed material subsequently. Near-ultra-violet tracer is followed by searching the beach or sea floor at night in the light of an ultra-violet lamp, or by taking samples of sand for examination under ultra-violet light in a darkroom. These fluorescent grains stand out clearly in ultra-violet light, and the method is a better means of detecting tracer at low concentrations than is a visual search for coloured grains. Russian geomorphologists who have traced sand movement on the Black Sea coast by the fluorescent tracer method have claimed that they can detect tracer at a dilution of one grain in 10 million of sand (Jolliffe, 1961).

Another method of tracing sediment flow depends on the use of radioactive tracers, which can be followed by means of a geiger counter or similar detection apparatus (Kidson and Carr, 1962). Artificial sand made from soda glass containing scandium oxide is ground down to the appropriate size and shape for use as tracer, then placed in a nuclear reactor to produce the radioactive isotope scandium 46. This has a half-life (i.e. time taken for the radioactivity to diminish to half its original strength) of eighty-five days, which means that it can still be detected three or four months after it has been dumped. Various other methods of preparing radioactive tracer have been developed, but the principle remains the same. After the tracer has been dumped, surveys are made by carrying detection apparatus over a beach, or dragging it across the sea floor mounted on a sledge. The location and intensity of radioactivity is mapped and paths of sediment flow can be deduced.

Radioactive tracers are expensive and require special preparation and careful use to avoid risks to public health, but they provide a good means of tracing sand movement, for tracer can be detected even when it has been buried beneath several inches of sand, where coloured or fluorescent tracer would not be visible. However, fluorescent tracers have advantages of cheapness and safety, and are often more durable in long-term experiments; radioactive tracer permits only one experiment at a time, but different colours of fluorescent dye can be used to trace movement from several points at the same time.

Mineralogical methods have also been used to trace sediment flow in and around the harbour being built at Portland, in W Victoria, in order to test the risk of the harbour entrance becoming choked with sand (Baker, 1956). After a preliminary survey of beach and sea-floor sands to determine the range of minerals present, quantities of six minerals not already present were introduced at different points, and the beaches in Portland Bay were sampled regularly to see where the tracer minerals appeared. It was found that the introduced grains moved across the floor of the bay, bypassing the harbour entrance, and arriving on the beaches to the E (Fig. 22); sediment flow on this part of the coast therefore appears unlikely to lead to sand accumulation in the harbour entrance.

The mineralogical method is laborious compared with radioactive or fluorescent tracer methods, because it depends on a complete preliminary inventory of the minerals present. The Portland experiment did not conform strictly with the requirements of a tracer experiment, for the introduced minerals were of higher specific gravity ('heavy minerals') than the coastal sediment, which consisted largely of quartz and shell sand. As lighter grains are moved more frequently and further than heavy grains, their long-term movement may not be in the same direction, and on some beaches, grains of different size, shape, and specific gravity are

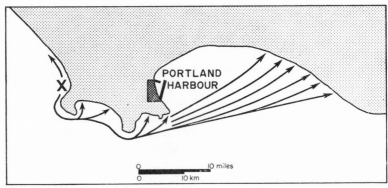

22   *Coastal sediment flow at Portland, Victoria, as determined by tracing material dumped at X (after Baker, 1956)*

moved in different directions over a long period. Differential sediment movement is well illustrated by the concentrations of heavy minerals left behind on beaches of the E coast of Australia by waves which have carried away the lighter quartz sand. It follows that the mineralogical method of tracing sediment flow is less reliable than the fluorescent or radioactive methods, and may in certain cases give misleading results.

All the methods so far described here are qualitative, designed to show where the coastal sediment goes, but in the last few years attempts have been made to measure rates of sediment flow. Yasso (1965b) used fluorescent tracers to determine rates of sand transport on the foreshore of Sandy Hook, New Jersey. Labelled sand in four size classes was injected, and sampling on a line 100 feet downdrift indicated that the smallest particles (0·59–0·70 mm) arrived first, and the next class (0·70–0·84 mm) soon afterwards, the maximum rates of flow being 2·6 and 2·0 feet per minute respectively. Once the predominant direction of sediment flow is known, it is possible to measure the volume of sediment moved by introducing standard quantities of tracer at regular intervals at a particular 'injection point', and taking samples at another point downdrift to measure the concentration of tracer passing by. The measured concentration of tracer is proportional to the rate at which the sediment is moving, so that quantities in transit can be calculated. The principle can be explained by analogy with a stream of water into which coloured dye is poured steadily at a particular point: the extent to which the dye is diluted by the water depends on the rate of flow of the stream, so that a high concentration of dye at a point downstream indicates a slow flow, and a low concentration a rapid flow; the volume of water moved over a particular period can be calculated from the rate of flow. The tracer concentration method has been applied successfully to the measurement of drift along shingle beaches on the S coast of England (Jolliffe, 1961), but its application to sand tracing is more difficult because of the vast number of grains involved.

Beach outlines in plan

Beaches often show smoothly-curved outlines in plan, concave seaward. It was formerly thought that these outlines were shaped by the action of strong currents sweeping sediment along the coast, but it is now realised that they are determined largely by wave action. Currents contribute to the movement of sediment to and fro along a beach, but they cannot shape depositional features built up above high tide level, and experiments in wave tanks have shown that smoothly-curved beaches can be produced by wave action alone (Lewis, 1938; Jennings, 1955).

On coasts exposed to waves generated by local winds but not to ocean swell, the outlines of beaches are related partly to the direction, strength, and frequency of onshore winds, partly to variations in the length of fetch or open water across which waves are generated by those winds, and partly to wave refraction. On the coasts of Denmark, Schou (1945) developed the idea that the orientation of beaches is the outcome of the long-term effects of wind-generated waves arriving from various directions. These effects he expressed as a resultant, calculated from records of the frequency and strength of winds, ignoring those of Beaufort Scale less than 4 (i.e. less than 21 km/hr) which have relatively little effect on beach outlines. A more precise relationship is obtained when the resultants are restricted to onshore winds, which are more important in producing the waves that shape coastal outlines than offshore winds. Similar resultants are used for the analysis of coastal dune orientations (see Chapter VI), the minimum wind speed for sand drifting being slightly less than Beaufort Scale 4 (about 16 km/hr). Vectors of wind forces are obtained by multiplying the frequency of winds in each Beaufort Scale class by the cube of the mean velocity of that class, and summing the results for each wind direction. They are expressed graphically with lines

of length proportional to the calculated vectors arranged in a
diagram, as in Fig. 23, and from this onshore resultants can be
determined for any coastline orientation.

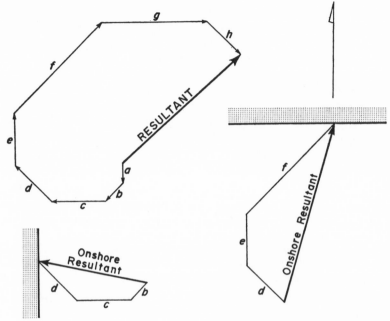

23   *Diagram to determine wind resultant (top left) from which onshore resultants
      are derived for an east-facing coast (bottom left) and a south-facing coast
      (right). a = vector line for N winds; b = vector line for NE winds, etc.*

Beaches built on coasts exempt from ocean swell, where there
are no great variations in fetch, tend to be modified by erosion
and accretion until they become orientated at right-angles to the
onshore wind resultant. Where there are marked contrasts in fetch,
however, these must be taken into account, for strong winds
blowing over a short fetch may be less effective than lighter winds
blowing over a long fetch. If the direction of longest fetch coincides
with the onshore wind resultant, the beach becomes orientated at
right-angles to this concident line, but where they differ the beach
becomes perpendicular to a line that lies between the two. On
coasts of intricate configuration, fetch becomes more important
than the onshore wind resultant, and beaches become orientated
at right-angles to the direction of maximum fetch (Schou, 1952).

Beach orientations on the coasts of the Baltic, the North Sea, and landlocked embayments or coastal lagoons, correspond well with these directional determinants. Local irregularities in the sea floor introduce a complication by refracting the waves approaching a shore so that beach outlines develop different alignments, cuspate behind offshore shoals and inset behind offshore hollows. Changes

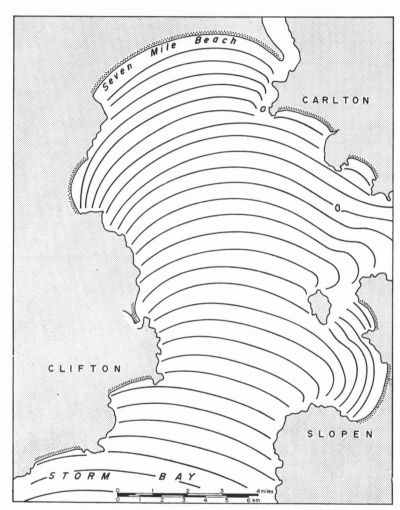

24 *Refracted wave patterns in Storm Bay, Tasmania, showing how the waves anticipate, then fit, the curved outlines of bordering sandy beaches (after Davies, 1959)*

in sea-floor configuration, where currents scour out hollows or build up shoals, change the patterns of wave refraction and lead to modifications of beach outlines (Robinson, 1966).

The shaping of beaches exposed to ocean swell is determined largely by patterns of wave refraction developed as the swell approaches the coast. Fetch is then too large to have any differential effect, and waves produced by onshore winds are of secondary importance in determining beach outlines. On calm days, a constructive swell is refracted into patterns that anticipate, then fit, the curved outlines of sandy beaches, each wave breaking synchronously along the shore (Plate 20). In Tasmania, Davies (1959) showed that southwesterly swell (wave period 14 seconds) becomes refracted to a southerly swell entering Frederick Henry Bay, and with further refraction develops outlines which fit the sandy beaches around that bay (Plate 1) but not the intervening sections of rocky shore (Fig. 24). Similar relationships exist between typical patterns of swell and the outlines of sandy beaches on other parts of coasts subject to ocean swell.

The effect of ocean swell is thus to modify the configuration of the coast, simplifying the irregular outlines produced by marine submergence into a succession of curved sandy shores between protruding rocky headlands, as on the coast of New South Wales (Fig. 48, p. 126). If the shoreline is less sharply curved than the approaching swell, lateral currents are generated which move sand to the centre of the bay, prograding it until it fits the outline of the arriving swell (Fig. 25). It is a process of mutual adjustment, for, as the sea floor is smoothed by erosion and deposition, wave refraction becomes less intense, and the outlines of beaches correspondingly less sharply curved.

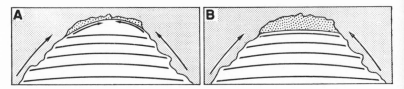

25   *Modification of the outline of a bay-head beach to conform with the refracted pattern of waves approaching the shore. In A, the outline of the beach is more sharply curved than the pattern of waves; longshore drifting (arrowed) carries beach material to the head of the bay until the outline of the beach has been adjusted to the wave pattern, as in B (after Davies, 1959).*

22 Shingle upper beach sharply demarcated from sandy lower beach with outflowing seepage channels on the E shore of Dungeness, England

23 Air photograph of part of Blakeney Point on the N Norfolk coast, England, showing shingle recurves surmounted by dunes, the outer shore in the background, and salt marshes with creek patterns (Cambridge University Collection)

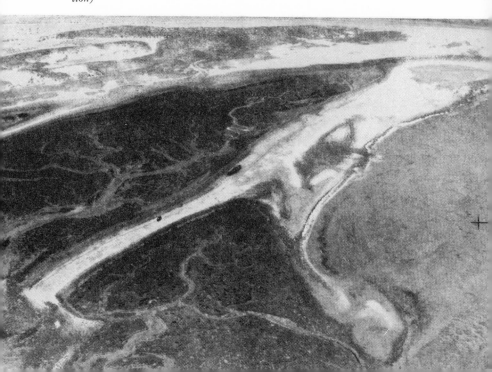

Attempts have been made to express the plan geometry of beach outlines mathematically. Yasso (1965a) examined the curvature of a number of North American beaches developed adjacent to headlands and subjected to a predominant direction of wave approach, and noted that the radius of curvature increased away from the headland in a curve reminiscent of the logarithmic spiral

$$r = e^{\theta \cot \alpha}$$

where r is the length of a radius vector from a log-spiral centre, θ is the compass bearing and α the constant-spiral angle between the radius vector and the tangent to the curve at a particular point. Computers were used to generate best-fitting log-spirals to mapped shoreline curves, and close relationships were obtained. Such a quantitative establishment of geometrical forms may permit a distinction between simple systems, where shoreline curvature is related to a predominant wave refraction pattern, and complex systems, where the outline is related to composite wave patterns, or where other complicating factors exist.

Beach outlines in profile

The profile or cross-section of a beach at any time is determined largely by wave conditions during the preceding period, and the effects of a severe storm may still be visible several months later. In calm weather, low waves form 'spilling' breakers with a constructive swash which moves sand or shingle on to a beach to build up a ridge or 'berm' parallel to the shoreline. In rough weather, higher and steeper waves form 'plunging' breakers, with collapsing crests which produce less swash, and a more destructive backwash which scours sediment away from the beach. According to Johnson (1956) waves with deep-water steepness (H_O/L_O) exceeding 0.25 are destructive, and those with smaller ratios constructive. The alternation of scour by storm waves and berm-building in calm weather is known as 'cut and fill'. Many sandy beaches show an eroded profile after storms, and a berm built up along the shore when calmer weather prevails (Davies, 1957). The effects of winter storms on a sandy beach at Apollo Bay, in Victoria, are shown by the successive profiles given in Fig. 26; the zone subject to removal and replacement is termed the sweep zone (King, 1959; King and Barnes, 1964). Measurements on the Ninety Mile Beach have shown that the seasonal sweep zone has a mean cross-sectional

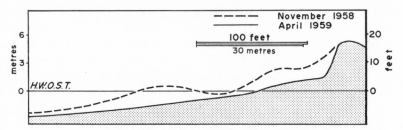

26 *Successive profiles measured on a sandy beach at Apollo Bay, Victoria, in November 1958, and after a storm in April 1959. The first profile showed a broad foredune and a beach berm, but in the second the foredune had been trimmed back and the beach had been lowered. This is typical of the alternation of cut and fill on sandy beaches: the zone between the pecked and the solid line is the sweep zone. H.W.O.S.T. = high water, ordinary spring tides.*

area of 42 sq.m, indicating a loss of 6 million m³ of sand from this beach during winter storms and equivalent replacements during fine weather, mainly in the summer months. Such a beach may be said to be in equilibrium in the sense that its profile variations are cyclically restored by natural processes over periods of time. Seaside resorts such as Deauville in France suffer when storms remove the beach that attracts their visitors; it may be replaced in calmer weather, but if the sand is carried too far offshore or swept along the shore beyond headlands it may not be returned by natural processes. In the United States, beaches lost in this way have been replenished artificially by pumping or dumping similar sand taken from coastal dunes or inland quarries on to the foreshore to restore the earlier cyclic equilibrium.

Beach profiles are modified by changes in the relative levels of land and sea. Other things being equal, submergence leads to recession of the beach and emergence to progradation, but it is necessary to take account of other factors, including variations in the incidence of cut and fill and the availability of beach sediment. A beach receiving abundant sediment may prograde even during a phase of submergence, while a beach that is losing sediment offshore or alongshore may be cut back even during a phase of emergence. Bruun (1962) suggested that if a beach profile has attained an equilibrium in relation to the processes at work on it, a relative rise of sea level will cause erosion of the upper beach and deposition in the nearshore zone in such a way as to displace the original profile landward. Schwartz (1967) has confirmed this, both from laboratory model experiments and from careful surveys of beaches near Cape Cod during the period from low spring tide to low neap tide, which is essentially a short-term sea level rise. It was found that the beach profile was displaced landward, erosion of the upper beach being compensated by nearshore deposition in such a way as to maintain the water depth adjacent to the shore (Fig. 27).

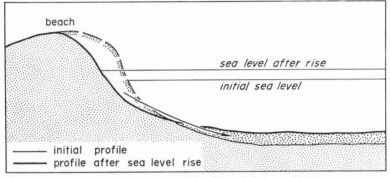

27   *Beach profile response to sea level rise (after Schwartz, 1967)*

Sand removed during storms is often retained as bars, awash at low tide, in the nearshore zone. These are concentrations of sand where the waves break, sand carried shorewards by the waves meeting sand withdrawn from the beach by the backwash. This effect has been reproduced in wave tanks, where it can be shown that the size of 'break-point bars' and their distance from the shore for a given calibre of sediments are related to the dimensions of waves: bigger waves build larger bars farther offshore (King and Williams, 1949). In calm weather, when constructive swash is more effective, bars move closer to the shore, and become flatter in profile as sand is passed into the swash zone and delivered to the beach face. Bars of this type are well developed off the Ninety Mile Beach, in Victoria, and at low tide the swell breaking on them forms a strong surf. They are interrupted by transverse channels formed and maintained by rip current systems, carrying the backwash seaward through the surf zone (Chapter II).

The alternation of cut and fill on a shore receiving plenty of sand may produce a succession of parallel ridges. Once a berm survives a storm, a new one is built up in front of it as sand is supplied to the beach during a succeeding phase of calm weather.

Prograded sandy beaches may show a series of parallel ridges, often surmounted by dunes built up in the manner described in the next chapter. At Woy Woy, to the north of Broken Bay, a large number of parallel ridges has been built up where cut and fill has influenced the mode of deposition of sand accumulating in a coastal embayment. The successive formation of ridges is marked by a sequence of soil and vegetation features, the most deeply-leached sand and the most advanced stage in vegetation succession being found on the inner ridges on the landward side, with a progression to little-leached soils and pioneer dune vegetation on the youngest ridges, towards the sea (Burges and Drover, 1953). Sandy ridges of this kind have also been described by Nossin (1965b) from the E coast of Malaya.

Parallel sandy ridges can also be formed by the successive addition of spits growing parallel to the shoreline. Successive maps of South Haven Peninsula on the Dorset coast indicate a historical evolution of this kind (Robinson, 1955), and a similar sequence can be seen on the W coast of the Falsterbo peninsula, in SW Sweden.

The profiles of shingle beaches differ in some ways from those of sandy beaches. This is partly because storm waves can have a slightly different effect: in addition to scouring shingle away from the beach face, the breaking waves throw some of it forward to build a ridge higher up the beach. In this way, a storm phase leads to the steepening of the beach profile. Subsequently, in calm weather, shingle returns to the beach face, restoring a gentler profile. Ridges on prograded shingle beaches are the outcome of successive storms, each of which threw up a ridge of shingle parallel to the shoreline (Lewis and Balchin, 1940).

It appears that shingle, and possibly also coarse sand (Thom, 1964), can be built into berms by storm wave action which is purely destructive on beaches of finer sand. On the N Norfolk coast, the storm surge of 1953 scoured away sandy beaches, but built up the crests of adjacent shingle beaches, sectors of which were rolled landward (Steers, 1953a). When the offshore profile is gentle, storm waves may break constructively, producing a strong swash, instead of plunging as they do on steeper shores. Under these conditions sand ridges can be built up along the shore by storm swash. Psuty (1965) has described parallel sand ridge formation by this process on the shores of the deltaic coast of Tabasco,

Mexico, where fluvially-supplied summer sand accretion is built into ridges by winter storm swash.

The height and spacing of parallel sand or shingle ridges are influenced by a number of factors, including the rate of supply of sand or shingle to the shore, the incidence of cut and fill, and changes in the relative levels of land and sea. A series of ridges showing an overall seaward descent in the levels of crests and swales may indicate progradation on an emerging coast, but the other factors must also be taken into account. Johnson (1919) discussed some of the difficulties encountered in attempting to use parallel ridge crest and swale levels as evidence of submergence or emergence of a coast.

On sandy shores where the tide range is sufficient to expose a broad foreshore at low tide, systems of ridges and troughs are found parallel, or at a slight angle, to the shoreline (Plate 21). These are known as 'low-and-ball' or 'ridge-and-runnel' in Britain where Gresswell (1953) has described them from tidal flats on the Lancashire coast. Their amplitude rarely exceeds 1 m, and the ridges are often as much as 100 m apart. As the tide falls the troughs may be temporarily occupied by lagoons, but these drain out by way of transverse channels as the ebb continues. Once established, these features persist through many tidal cycles; they are associated with wave action as the tide rises and falls, for when the waves have been arriving at an angle to the shore the ridge pattern also runs obliquely. The transverse channels are related to rip currents in the nearshore zone. On the Australian coast these features occur where the tide range is large, particularly where the effects of ocean swell are weak or excluded: the author has observed them at low tide on the shores of Westernport Bay in Victoria, and on the shores bordering Gulf St Vincent, in South Australia, and similar features have been described by Russell and McIntyre (1966) from tidal flats in the Darwin area and the coast near Port Hedland in N Australia.

On a smaller scale, a variety of rippled surfaces may develop on the lower beach or foreshore exposed at low tide, produced by bottom currents associated with wind, wave, and tidal action. Simple transverse current ripples are typically asymmetrical, with a steeper face away from the current, and associations of cross-currents, generated by winds, waves, and tides, can produce intricate networks of transverse current ripples. In addition, there are longitudinal ripples, aligned parallel to a strong unidirectional current, and these also can be complicated by interfering wave motion and cross-current patterns (Van Straaten, 1953). Tanner (1963) devised mechanical methods of simulating ripple-producing bottom-shear, and found that eddy currents are important in steepening and sharpening subaqueous ripples to produce the forms observed on tidal foreshores.

Features related to sorting

Sorting of sediment on the beach by wave action takes several forms. Concentrations of heavy minerals are often found at the back of a beach after storms when waves have carried away the lighter sand grains, and seams of heavy minerals deposited in this way may afterwards be discovered beneath dunes built up behind the beach. Stratified layers of finer and coarser sediment within a beach record alternations of stronger and weaker wave action. Superficial sorting often produces zones of shingle and coarse and fine sand parallel to the shoreline, sometimes with a steeper upper beach separated by a break of slope, often a seepage line, from a flatter lower beach exposed as the tide recedes. On the English coast the upper beach is often of shingle, clearly demarcated from the lower beach of sand, the contrast probably reflecting the different response of shingle and sand to storm wave action, which tends to pile up a shingle ridge and to withdraw the sand (Plate 22). Within each stratified layer or beach face zone the distinctive sediment is well sorted, being related to a particular phase of deposition or reworking by waves.

Lateral gradation in mean particle size is also common along beaches, notably on Chesil Beach, which grades from granules and coarse sand on the beach crest at the W end near Bridport through to pebbles and cobbles on the higher (12 m above high tide level) beach crest at the E end near Portland Bill (Fig. 28). This gradation may be related to a lateral increase in wave energy, with higher waves coming in through deeper water to build a higher and coarser beach at the more exposed SE end. Bascom (1951) related a similar gradation in beach sand particle size at Half Moon Bay, California, to variation in degree of exposure to refracted ocean swell coming in from the NW, coarse sand on the exposed sector grading to fine sand in the lee of a headland at the northern end. Lateral gradation of particle size may alternatively

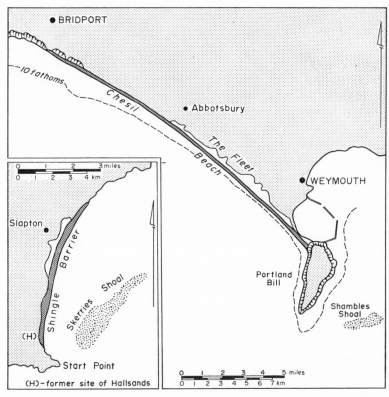

28　*Chesil Beach, a shingle barrier on the Dorset coast and (inset) the shingle
barrier and lagoon at Slapton, in Devonshire, England*

be the outcome of wearing, attrition, and selective sorting of shingle
derived from a particular sector of the coast as longshore drifting
carries it away from the source area. Marshall (1929) recorded an
example of this on the shores of Hawke Bay, New Zealand, in a
sector fed by gravels drifting from the mouth of Mohaka River,
and Landon (1930) described how angular gravel from a cliff on
the W coast of Lake Michigan becomes rounded as it drifts
southward along the shore.

A different kind of sorting produces 'beach cusps' in the swash
zone on sand and shingle beaches (Russell and McIntyre, 1965a).
These consist of regular successions of half-saucer depressions,
up to 1 m deep between cusps of slightly coarser material up to
30 m apart. They are ephemeral features, but once formed they

influence the patterns of swash as the waves break. Their mode of formation is not fully understood; it has been shown that larger waves build bigger cusps, and they are most frequently found where the waves arrive parallel to the shoreline, rather than obliquely. They are also best developed where there is a less permeable layer beneath the superficial beach deposits (Longuet-Higgins and Parkin, 1962). In Australia they frequently form on steep coarse sandy beaches as in the bays of the Sydney coast and also on the sandy shores of landlocked embayments, such as Botany Bay.

Wave-built beach forms are essentially similar on coasts in various climatic environments, except in high latitudes where cold conditions introduce important modifications. When the sea is frozen, wave and nearshore processes are halted, but when storm waves break up a winter ice fringe and drive it on to the shore, ice-pushed mounds and ridges are formed. These have been described from Arctic coast beaches in Alaska, N Canada, and Siberia, and from beaches in Antarctica, where ice-rafted erratic rock fragments and the debris yield of coastal glaciers are added to the beach, and summer meltwater cuts channels across the back-shore. Beach gravels are less well rounded than on shingle beaches in warmer regions, the addition of frost-shattered and solifluction gravel outweighing the rounding and smoothing effects of summer wave action. The beaches are also less well sorted and often include muddy zones. Freeze-and-thaw processes form cracks and mounds and produce stone polygons on the poorly sorted beach material. Yet the characteristic beach forms, including beach ridges, spits, and tombolos, have developed on the N and W coasts of Alaska, and Nichols (1961) has described them from the ice-free shores of McMurdo Sound in Antarctica.

Beach rock

Beach rock is formed where a layer of beach sand becomes consolidated by secondary deposition of calcium carbonate at about the level of the water-table (Russell, 1962). The cementing material is precipitated from ground water in the zone between high and low tide level, which is subject to repeated wetting and drying as the water-table rises and falls with the tide, or during and after wet weather. Some believe that precipitation of calcium carbonate is aided or brought about by the action of micro-organisms, such as bacteria, which inhabit the beach close to the water-table. Often the cementing material is aragonite, probably derived from sea water, rather than calcite, derived from ground water (Stoddart and Cann, 1965).

Beach rock has been found on the shores of the Caribbean Sea, around the Mediterranean, the Red Sea, and the Persian Gulf, and on the coasts of Brazil, South Africa, and Australia; it is commonly found on sand cays built up on reefs off tropical coasts (Plate 32, p. 175). Its restriction to warm environments has been attributed to a requirement that interstitial water should have a temperature exceeding 20°C for at least half the year in order to produce cementing of the beach sediment (Russell and McIntyre, 1965b). As it forms at a definite horizon in relation to sea level, the occurrence of beach rock above this level is a good indication that emergence has taken place. Relatively resistant, beach rock may impede shoreline erosion; it often persists in patterns which commemorate the former outlines of beaches and cays that have since been eroded away.

Spits

Spits are depositional features built along the shore, usually ending in one or more landward hooks or recurves. They were formerly ascribed to current action, but although currents may contribute sediment to them, they grow in the predominant direction of longshore sediment flow caused by waves, and their outlines are shaped largely by wave action. The recurves are formed either by the interplay of sets of waves arriving from different directions (Fig. 29), or by wave refraction around their distal ends (Evans, 1942). Traces of older recurves on the landward side mark former terminations of a spit prolonged intermittently, a feature well shown by Cape Bowling Green (Fig. 80, p. 183) on the N Queensland coast. Salt marshes often develop on the sheltered landward side of spits, notably between recurves, as on Blakeney Point (Fig. 30 and Plate 23) and Scolt Head Island in Norfolk, where spits have been driven landward athwart the salt marshes, so that marsh deposits outcrop on the seaward side (Fig. 31). Other spits have been widened by the addition of successive ridges on the seaward side, and stages in their evolution can be deduced from the pattern of beach ridges, as on the spit in Carrickfergus Bay, on the E coast

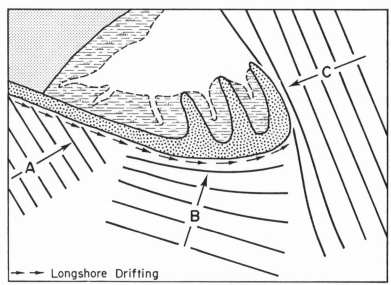

29   *The shaping of a recurved spit (based on the outline of Hurst Castle Spit, S England): waves from A, arriving at an angle to the shore, set up longshore drifting which supplies sediment to the spit; waves from B and C determine the orientation of its seaward margin and recurved laterals respectively*

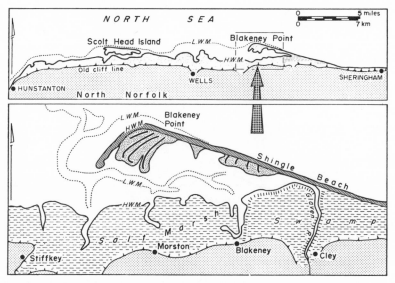

30   *Blakeney Point, a recurved spit on the N Norfolk coast*

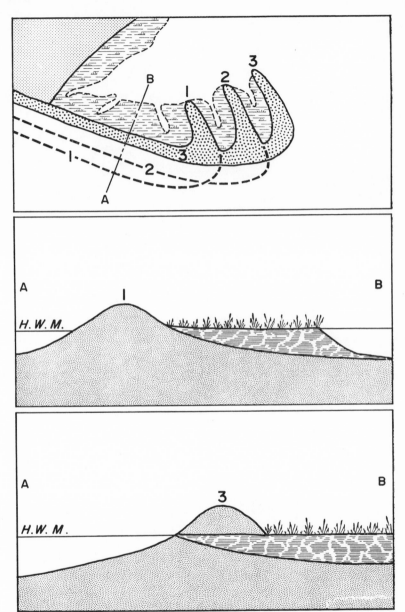

31   *Stages in the evolution of a recurved spit. The upper diagram shows three stages
      in development, with successive terminal recurves, and marshland formed on
      the sheltered side. The middle and lower diagrams are sections along AB at
      the first and third stages, showing how the main bank has been driven land-
      ward over the marshland until marsh deposits outcrop on the seaward side.*

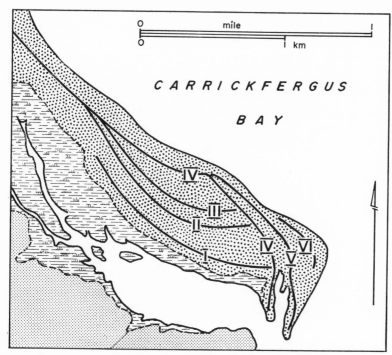

32 *The recurved spit in Carrickfergus Bay, Tasmania. Selected beach ridges (I–VI) show former outlines of the shore, stages IV, V, and VI having been added after an earlier spit (stages I–III) had been truncated by erosion (after Davies, 1959).*

of Tasmania (Fig. 32). Sand eroded from cliffs of glacial drift on the Cape Cod peninsula has been built into a spit, prograded on the seaward side, at Provincetown (Fig. 33). As Davis (1896) showed, the fulcrum of this spit has migrated up the coast so that part of the formerly prograded sector has now been truncated by marine erosion. The recurved spit at Sandy Hook, New Jersey, originated in a similar way. The evolution of spits such as Blakeney Point and Orford Ness on the East Anglian coast can sometimes be traced from old maps showing earlier stages in their growth, bearing in mind the limitations of historical cartographic accuracy noted by Carr (1962). The shingle spit extending from Orford Ness has deflected the mouth of the River Alde about 11 miles south, and maps made since about A.D. 1530 suggest that it has grown about 4 miles during the past four centuries (Steers, 1953a).

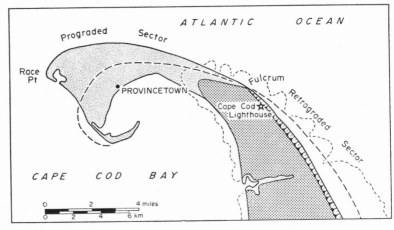

33   *The outline of Cape Cod, Massachusetts, U.S.A.*

This kind of information is rarely available outside W Europe and North America, but eventually it will be possible to use successive surveys and series of dated aerial photographs to trace the evolution of spits. Carr (1965) used this kind of evidence to analyse short-term changes at the S end of the Orford Ness spit over the period 1945-62.

Some of the best examples of spits are found on the shores of landlocked seas, lakes, and coastal lagoons, where sand and shingle carried along the shoreline are deposited as spits where the orientation of the shoreline changes, in forms related to prevailing wave conditions. Examples of this are found in the Gippsland Lakes, where the recurved sand and shingle spit at Butlers Point has been shaped by waves arriving from relatively long fetches to the E, SE, and NW, while at Metung a shingle arrow-spit has grown out from a promontory at a point of convergence of longshore drifting of beach gravel (Fig. 34). These spits show on a small scale forms comparable with the sand spits in the Danish archipelago (Schou, 1945). Arrow-spits also develop in the lee of islands, as in the 'comet-tails' off the Brittany coast, and where the island has been eroded away the depositional trail may persist as a 'flying spit', either elongated in the direction of sediment flow or aligned at right-angles to the predominant waves, with terminal recurves at each end. Examples of these forms occur on the Boston coast

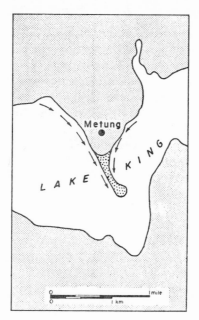

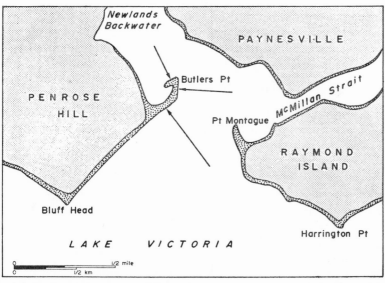

34   *Spits on the shore of the Gippsland Lakes, coastal lagoons in SE Australia*
     (cf. Fig. 47). The recurved spit at Butlers Point has been shaped by waves
     arriving from three directions (cf. Fig. 29) and the arrow-spit at Metung
     marks a sector of convergence of longshore drifting of sand and shingle.

(Nichols, 1948) and in the Strait of Georgia, W Canada, in both cases derived from islands of glacial drift.

Other depositional forms include tombolos, which are spits linking an island to the mainland. The term comes from Italy, where these features are well developed, and Johnson (1919) has described examples from New England. In Australia the Yanakie isthmus which ties the granitic upland of Wilsons Promontory to the coast of Victoria, is a tombolo, as is the sand isthmus at Palm Beach, linking Barrenjoey to the mainland (Fig. 35). Cuspate spits

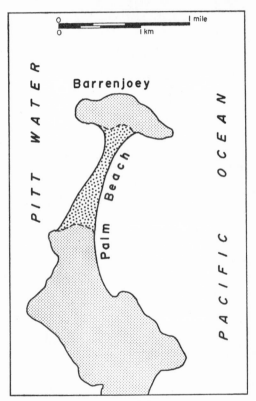

35   *The tombolo at Palm Beach, New South Wales, linking Barrenjoey Island to the mainland*

form in the lee of shoals or offshore islands, and are shaped by waves refracted round the island (Gabo Island, Fig. 36); they may eventually grow into tombolos linking the island, as at Cape Verde

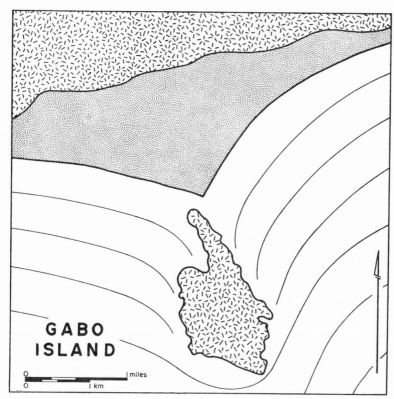

GABO
ISLAND

36 *Cuspate sandy foreland developed in the lee of an offshore island, around which waves are refracted, at Gabo Island, SE Australia*

on the Senegal coast. Cuspate spits also form on the shores of lagoons and narrow straits, where waves approaching from each diagonal are stronger and more frequent than those coming directly onshore, so that, at points of convergence of longshore drift, sediment is built up from two directions (see p. 170). Similar features may develop where currents converge, or in relation to an eddy in a longshore current: a possible explanation for Cape Hatteras and the other cuspate spits on the E coast of the United States, although it has been suggested alternatively that these relate to wave refraction by offshore shoals (Russell, 1958). Cuspate forelands are spits that have been enlarged by the accretion of ridges parallel to their shores, and stages in their evolution may often be deduced from the patterns of these ridges. Some have remained stationary, and developed symmetrically, while others have migrated, so that one set of ridges has been truncated, and is bordered by the other set (Fig. 37). Dungeness, a cuspate shingle foreland on the Channel coast of England (Fig. 38) has a ridge pattern indicative of eastward migration during its accretion

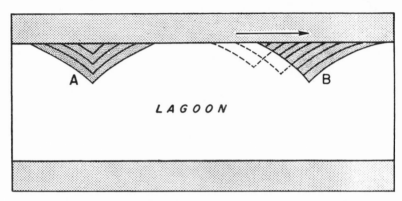

37 *Cuspate forelands on a lagoon shore. Beach ridge patterns indicate that A has grown symmetrically, remaining in position as accretion of sediment took place, whereas B has migrated along the shore, one set of ridges being truncated by the other.*

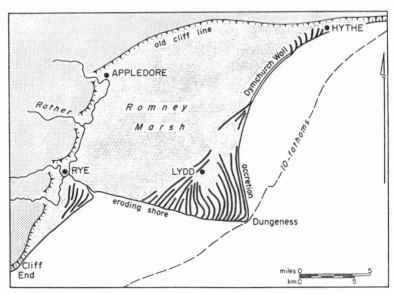

38 *The cuspate shingle foreland at Dungeness, England (after Lewis, 1932)*

(Lewis, 1932). Cape Kennedy and the Darss peninsula on the Baltic coast of East Germany are other examples of cuspate sandy forelands of complex origin (Johnson, 1919).

River mouths and lagoon entrances are often deflected by the growth of spits, prolonged in the direction of longshore drifting (Fig. 39). As the spit grows, the transverse ebb and flow of tidal

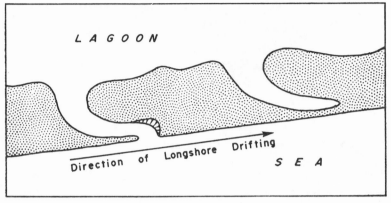

39　*Deflection of lagoon outlets in the direction of longshore drifting*

currents is impacted against the farther shore which is cut back by current scour. Some deflected entrances become sealed off altogether, leaving a channel that formerly led to an outlet. Migrating entrances of this kind punctuate the sandy outer barrier on the Atlantic coast of the United States: the Outer Banks of North Carolina have had a long history of entrance deflection, migration, and closure (Dunbar, 1956). Non-migrating entrances are often bordered by paired spits, with outlines influenced by patterns of waves, refracted as they enter. These may indicate the constriction of a tidal entrance by convergent longshore drifting, but they can also develop alongside an entrance breached through a coastal barrier (Kidson, 1963). Spits of this kind border the entrances to estuarine harbours on the S coast of England at Poole, Christchurch, and Pagham (Robinson, 1955), and are well developed on the flanks of entrances to Corner Inlet, in Victoria (Fig. 40). The Ninety Mile Beach is here being prolonged by a sand spit growing southwestwards, supplied by beach material that drifts down the E Gippsland coast when easterly waves approach the beach. A spit of this kind, prolonged until it stands in front of a former coastline, or seals off inlets and embayments, becomes a coastal barrier.

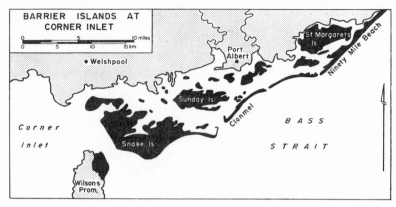

40   *Sandy barrier islands at Corner Inlet, Victoria*

Barrier beaches and related forms

The deposition of beach material offshore, or across the mouths of inlets or embayments, in such a way as to form barriers extending above the normal level of highest tides and partly or wholly enclosing lagoons, is a widely-distributed phenomenon which has received considerable attention in recent years. Barriers, thus defined, are distinct from bars, which are submerged for at least part of the tidal cycle, and from reefs of biogenic origin, built by coral and associated organisms (Chapter IX). They show a variety of forms. *Barrier beaches* are narrow strips of low-lying depositional land consisting entirely of beach sediment, without surmounting dunes or associated swamps. Many barriers do have these additional features, and some attain widths of several kilometres, with crests of dunes sometimes rising more than a hundred metres above sea level. The term *bay barrier* describes a feature built up across an embayment, and *barrier island* indicates a discrete segment, often recurved at both ends. Fig. 40 shows barrier islands and the spit at the SW end of the Ninety Mile Beach, and Fig. 41 shows part of the Frisian Islands, a barrier island chain with intervening tidal entrances on the German North Sea coast. Chains of barrier islands separated by tidal entrances are also well developed on the Gulf and Atlantic coasts of the United States, and on sectors of the African, South American, Indian, Russian, and Australian coasts, as well as on a smaller scale elsewhere. It is clear from recent work that barriers have originated in a variety of ways, and that no single explanation will account for all these depositional features.

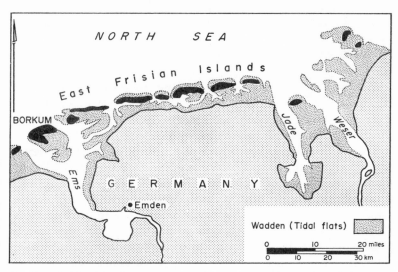

41 *East Frisian Islands, a barrier island chain on the German North Sea coast*

The two main modes of barrier origin are by longshore growth of spits (Fig. 42) and by development of emergent beaches offshore, but many barriers have had a composite origin. On the West African coast, Guilcher and Nicolas (1954) have described elongated barrier spits such as the Langue de Barbarie extending as the result of southward longshore drifting, in such a way as to deflect estuaries, notably the mouth of the Senegal. On the other hand, the barrier islands bordering the Texas coast have the appearance of features that were initiated offshore. Johnson concluded that barriers of this kind were most likely to form on coasts bordered by shallow seas as the result of a relatively sudden emergence, when reduction of water depth offshore caused waves to break farther out, building a bar which emerged as it was driven shoreward. It is possible to generate this sequence of events experimentally in a wave tank, but it is doubtful if it applies to many actual barriers, for the majority of these came into existence during the episode of submergence marked by the Holocene marine transgression (Russell, 1958), and many of them are the outcome

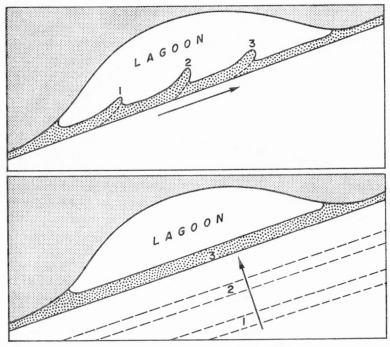

42   *Stages in the evolution of a barrier to enclose a lagoon: by prolongation of a
      spit (above) or by shoreward migration of a barrier that originated offshore
      (below)*

of shoreward sweeping of sediment during, and perhaps also since, this marine transgression.

   This is well illustrated by the shingle barrier at Chesil Beach, on the English coast, which stands in front of a lagoon (the Fleet), and an embayed mainland coast which has escaped marine cliffing (Fig. 28). Evidently this barrier came into existence during the marine transgression in such a way as to protect the submerging mainland coast. Steers (1953a) suggested that the shingle barrier developed during the earlier stages of the Holocene transgression, when the rising sea collected gravels that had previously been strewn across the emerged sea floor, and swept them shoreward. Its origin as an emerged feature still presents a problem. It is possible that it first developed during one of the sea level regressions that occurred during the oscillating Holocene transgression, and was thereafter driven shoreward as a barrier. Chesil Beach is still moving shoreward intermittently, for vigorous storms sweep shingle over the crest and down into the waters of the Fleet. Eventually it will come to rest against the mainland coast, which will thereafter be cut back as marine cliffs by storm wave activity: rejuvenation of the coastal slopes is spreading eastwards from Bridport as this evolution proceeds. Zenkovich (1967) has invoked a similar sequence of events to explain the barriers of sand and shingle that have developed in front of uncliffed mainland coasts in the Soviet Union, and Le Bourdiec (1958) reached a similar conclusion for barriers on the Ivory Coast. It has also been suggested that shoreward sweeping of sediment during the Holocene marine transgression initiated the barrier island system which borders the Gulf coast of the United States (Shepard, 1960, 1963). In each case the barrier formations rest upon an older (Pleistocene) land surface submerged by the Holocene transgression (Fig. 43).

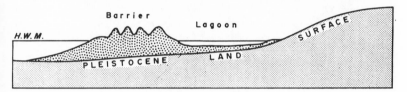

43   *Section through a barrier enclosing a lagoon*

The supply of sediment for the building of barrier systems raises the same problems as were previously considered in relation to beach nourishment. Chesil Beach is apparently receiving little new shingle either from onshore or longshore drifting at the present time, and experiments with radioactive pebbles dumped on the sea floor failed to show any clear evidence of shoreward drifting to Scolt Head Island, a barrier island on the N Norfolk coast, under present conditions. These features are evidently relict formations, a legacy of the Holocene marine transgression.

Other barrier systems are still receiving sediment derived from adjacent cliffed coasts, river discharge, and possibly shoreward drifting. The barriers which extend across embayments on the New England coast, enclosing lagoons or ponds, are built of sediment derived from adjacent cliffs cut in glacial outwash material. They are well developed on the S shores of Martha's Vineyard (Fig. 44) and Nantucket Islands off the Massachusetts coast, and on the S shores of Cape Cod (Johnson, 1925). The barrier

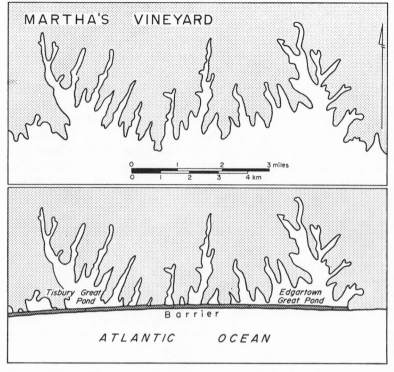

44   *Barriers and lagoons at Martha's Vineyard Island, Massachusetts, U.S.A.*
     (after Johnson, 1925)

spits termed *nehrungen* on the S Baltic coast are similarly derived
from glacial deposits eroded by the sea and carried along the shore
to enclose embayments as lagoons. Vladimirov (1961) described
barriers and spits derived from intervening cliffs of Tertiary
sediment on the coast of Sakhalin. Predominantly fluvial nourishment is indicated by King (1959) for barriers in S Iceland built
by the constructive action of Atlantic swell working on the
superabundant sediment, chiefly sand, supplied to the coast by
glacifluvial streams. The sand and shingle barrier enclosing Lake
Ellesmere on the New Zealand coast (Fig. 45) is built largely of
reworked glacifluvial material supplied by the Rakaia and other
rivers draining the Canterbury outwash plains.
    On the Australian coast there are multiple barrier systems, with
intervening tracts of lagoon and swamp. The inner barriers are

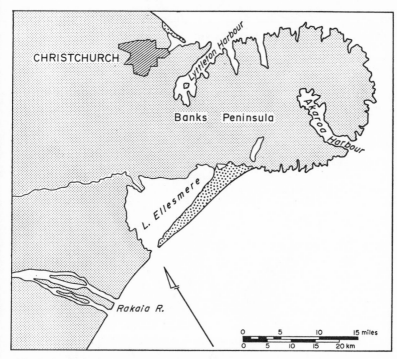

45 *Lake Ellesmere, a coastal lagoon near Banks Peninsula, South Island, New Zealand*

generally of Pleistocene age, having been dissected by stream incision during the Last Glacial (and perhaps earlier) low sea level phases, whereas the outer barrier is often of Recent age, having come into existence during and since the Holocene marine transgression. In detail there are many complications, for barriers have been built, then dissected or destroyed, and subsequently rebuilt, sometimes in overlapping alignments, during the Quaternary episodes of sea level oscillation. Complex barrier systems have been investigated on the Victorian coast (Bird, 1965) (Fig. 47) and on the coast of New South Wales (Thom, 1965) (Fig. 48). These have developed in predominantly quartzose sand, and have not become lithified in the manner of the calcareous sand barriers of S and W Australia, which are commonly preserved as relatively durable calcarenite. In the SE of South Australia there are numerous parallel ridges of calcarenite which formed as successive barriers during Pleistocene times. The sequence is indicative of intermittent

coastal emergence, due in part to tectonic uplift accompanying
eustatic changes. The intervening tracts are corridors of swamp

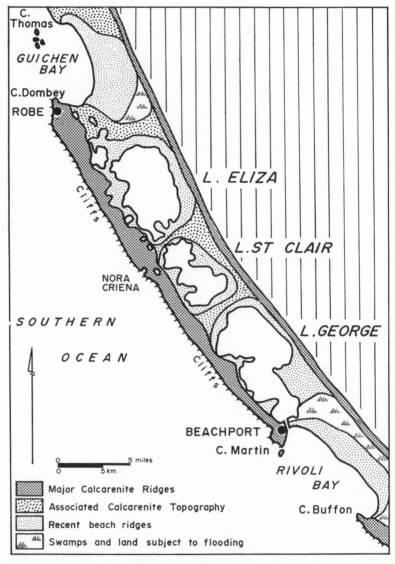

46   *Coastal lagoons between Robe and Beachport, South Australia, between an*
     inner barrier of calcarenite and a similar outer barrier which is undergoing
     dissection and denudation on the seaward side

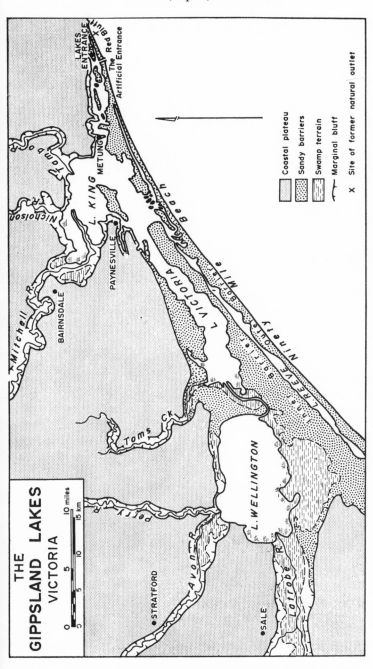

47 *The Gippsland Lakes, a group of coastal lagoons in SE Australia enclosed within a former embayment by a series of coastal barriers*

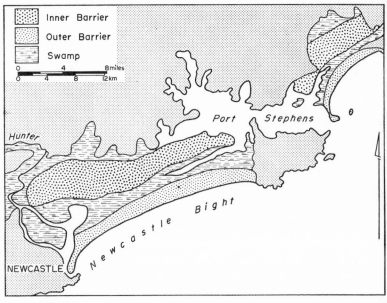

48 *Coastal features near Newcastle, New South Wales, where inner and outer barriers have developed successively in a coastal embayment (after Thom, 1965)*

land, developed in former lagoons. The Coorong, behind the outer barrier on the shores of Encounter Bay, is the last in the sequence of inter-barrier lagoons, not yet filled by swamp deposits. Complexity of barrier evolution is illustrated farther south, between Robe and Beachport (Fig. 46), where the coast consists of a cliffed and dissected ridge of calcarenite, probably of Pleistocene age, there being no Recent outer barrier of unconsolidated sand here.

The sandy barriers on the SE coast of Australia are not migrating shoreward in the manner of the shingle barrier, Chesil Beach, except to a local and limited extent where winds or storm waves carry sand over into the lagoon. Many of these barriers have been prograded by the addition of successive parallel beach ridges on alignments which, like the curved outline of the present beach, result from the refracted patterns of dominant swell approaching the coast. Often the beach ridges are surmounted by parallel dunes, developed successively during progradation. The pattern of beach ridges and dunes can be used to decipher the history of barrier evolution, particularly where there are relics of former recurves indicative of stages in longshore growth. The outer barrier on the E Gippsland coast (Fig. 47) originated as a chain of barrier islands which were extended both NE and SW by longshore drifting, deflecting outlets from coastal lagoons and in some cases sealing off the deflected channels (cf. Fig. 39). Subsequently, this barrier was widened by progradation, with successive parallel ridges added on the seaward side to take up the alignment of the Ninety Mile Beach along the present shore (Bird, 1965). By contrast, the Outer Banks of Carolina N of Cape Hatteras are migrating landward as the result of spillover dunes and washover sand invading Pamlico Sound and gradual recession, rather than progradation, of the seaward margin. Transgressive barrier formations result in complex patterns of overlapping stratigraphy on the Dutch coast (Van Straaten, 1965).

Barriers are best developed where the tide range is relatively low, as on the SE coast of Australia. A large tide range generates strong ebb and flow currents which maintain gaps through barriers and prevent wave action from depositing sand or shingle to seal them off. On the E Gippsland coast the spring tide range increases from about 1 m at the NE end, where there are no natural tidal entrances, to 2·5 m at the SW end, where tidal currents have maintained channels between a chain of barrier islands at the entrance to Corner Inlet. As a result, the Gippsland Lakes (Fig. 47) were amost completely sealed off from the sea at the eastern end while Corner Inlet, a similar embayment, has an incomplete barrier system (Fig. 40). A similar contrast exists between the relatively unbroken *nehrungen* on the almost tideless S coast of the Baltic Sea and the chain of barrier islands with intervening tidal entrances on the strongly tidal S coast of the North Sea (Fig. 41).

It is clear that the factors and processes involved in the origin and shaping of depositional beach, spit, and barrier formations are many and varied, and it is difficult at this stage in our knowledge of these features to propound typical modes of evolution. This is a field in coastal geomorphology that requires many more detailed and long-term studies of the form and dynamics of particular examples.

VI
COASTAL DUNES

Coastal dunes are formed where sand deposited on the shore dries out and is blown to the back of the beach. Where the tide range is large, as on the Atlantic coast of Devon and Cornwall, sand blown from broad foreshores exposed at low tide is built up as dune topography extending inland from high tide mark, as at Braunton Burrows, Perran Sands, and Gwithian Sands. Dunes are similarly derived from broad inter-tidal foreshores on the North Sea coasts of Belgium, Holland, Germany, and Denmark. On coasts where the tide range is small, sand delivered to the beach by wave action may provide sufficient material for dune construction, as on the W and SE coasts of Australia, the W coast of Africa, and sectors of the Atlantic and Pacific coasts of the United States.

On some coasts wind-blown sand has been accumulating during and since Pleistocene times, producing extensive and complicated sequences of dune topography. Where the parent sands are calcareous, as on the W coast of Australia, the older dunes have been lithified by internal deposition of calcium carbonate from percolating water, forming dune limestone, an aeolian calcarenite which preserves the dune topography in solid rock, but where the parent sands are quartzose, as on Australia's SE coast, this kind of lithification has not taken place, and the dunes remain unconsolidated—either active and mobile, or retained by a vegetation cover. In South Australia there is a lateral transition in the composition of dune sands SE of the Lower Murray, the proportion of terrigenous quartzose sand diminishing as the proportion of marine calcareous sand increases southeastwards (see p. 88). The strongly calcareous dune ridges to the SE are preserved in solid calcarenite, but they grade northwestwards into more irregular and disrupted dune topography, lithification diminishing as the sand becomes less calcareous. On the E Mediterranean coast there is a similar transition, the calcarenite dune topography of the Israel coast passing southward into more irregular, mobile dunes on the coast of the Gaza Strip, where the sands have an increasing content of terrigenous (Nile-derived) sediment.

Coastal dunes are best developed on coasts in the temperate and arid tropical zones: in the humid tropics they are of limited and local extent (Jennings, 1964). Even where sand has been delivered to the shore, as in E Malaya (Nossin, 1965b), the coastal topography consists of low beach ridges with little or no dune development. It has been suggested that the prevailing dampness of beach sands on humid tropical coasts impedes deflation and backshore dune development, particularly on microtidal sectors where the beaches are narrow. This may be a contributory factor, but observations on temperate coasts indicate that strong onshore winds can dry off a beach surface and blow away sand even when it is actually raining. Another suggestion is that rapid colonisation of coastal sand accumulations by luxuriant vegetation prevents dune development in the humid tropics, but it is doubtful if the vegetation has much effect: if aeolian sand were arriving at the backshore, dense vegetation would simply trap it and facilitate the building of a high foredune. It seems to be a question of failure of sand supply. As Jennings (1965) has pointed out, strong winds are rare in the humid tropics in comparison with other climatic zones, and the occasional violence of tropical storms is usually accompanied by torrential rainfall which saturates the beach surface and impedes sand transportation by the temporarily strong wind action.

Characteristics of dune topography have been studied in various parts of the world. Notable contributors have been Van Dieren (1934) on the Dutch coast, Briquet (1923) in France, Schou (1945) in Denmark, Steers (1964) in Britain, Cooper (1958) in North America, and Guilcher (1958a) in West Africa. The following account is mainly in terms of dune topography as developed on the Australian coast, with incidental references to other areas.

Foredunes

Foredunes are built up at the back of a beach or on the crest of a beach ridge of sand or shingle where dune grasses colonise and start to trap blown sand (Plate 24). They become higher and wider as accretion continues. There are regional variations in the grass species which act as pioneer colonists and foredune-builders. In the British Isles and W Europe the typical pioneer dune plant is marram grass *(Ammophila arenaria)*, a species that has been introduced to many other parts of the world, including Australia, as a dune stabiliser; lyme grass *(Elymus arenarius)* and the sea

24 *New foredune built by sand accretion in marram grass on a berm, Ninety Mile Beach, Lakes Entrance, Victoria*

25 *Parabolic dunes advancing inland from the Queensland coast near Cape Flattery*

wheat grass *(Agropyron junceum)* are also common dune pioneers in Europe, the latter extending around the Mediterranean and Black Seas. In Australia and New Zealand the native dune pioneers are coast fescue *(Festuca littoralis)* and sand spinifex *(Spinifex hirsutus)*. In the United States another form of marram grass, *Ammophila brevigulata*, is a common pioneer species. Marram grass is the most vigorous of the dune-building plants, thriving where it is able to trap blown sand, and capable of adding several feet of sand to a developing foredune in the course of a single year. However, where sand accretion is slow, marram grass grows poorly, and other grasses often take its place.

On some coasts a single foredune persists, cut back from time to time by storm waves on its seaward side but restored subsequently by accretion of wind-blown sand in recolonising vegetation. A dune of this kind borders the NE coast of Norfolk in the vicinity of Sea Palling, where it has revived after severe damage and breaching during the 1953 storm surge. But many coasts have dunes arranged in patterns of ridges parallel to the shoreline, formed by the development of successive foredunes on a prograding sandy shore. These are particularly well developed in SE Australia, notably on the coastal barriers of E Gippsland, but there are similar features on many other coasts, including parts of the E coast of Britain, at Tentsmuir in Scotland, and Winterton Ness in East Anglia. The Australian examples are believed to result from an extension of the cut and fill process described in the previous chapter, when the resulting beach ridges become the foundations for successive foredune development (Fig. 49).

Parallel dunes

When the seaward margin of a foredune is trimmed back by waves during a storm, a crumbling cliff of sand is exposed (Fig. 49c). Subsequently, during calmer weather, waves build up a new beach ridge in front of, and parallel to, the trimmed margin of the foredune, separated from it by a low-lying trough or swale (Fig. 49d). Dune grasses tend to colonise the new beach ridge first, leaving the swale unvegetated, so that sand accretion is concentrated along the line of the beach ridge, and a new foredune is built up. Continued growth of the new foredune gradually cuts off the supply of sand to its predecessor, which becomes relatively stable, and the dune grasses are then invaded and replaced by scrub

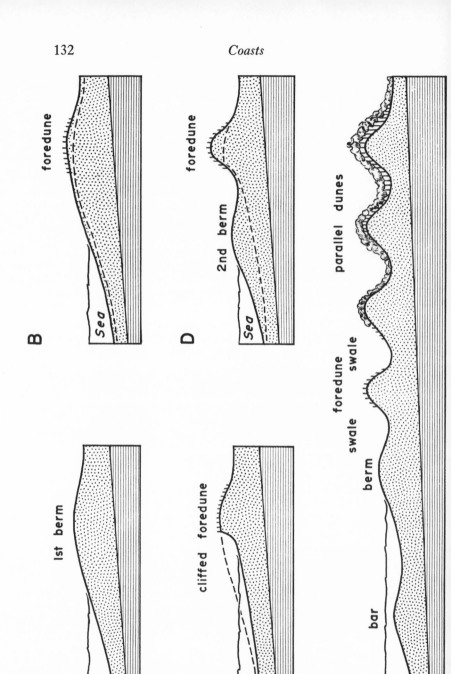

vegetation, which in South Australia is dominated by the coast tea-tree *(Leptospermum laevigatum)*, and in Britain by a rose and bramble thicket with buckthorn scrub *(Hippophae rhamnoides)*.

If shoreline progradation continues, alternating episodes of cut and fill lead to the formation of a series of parallel dunes, separated by swales (Fig. 49e), and as each foredune is built upon a beach ridge formed by constructive waves the alignment of the parallel dunes is determined indirectly by the characteristic patterns of refracted ocean swell (Plate 20). The process of parallel dune formation is well illustrated on the coast at Lakes Entrance, in Victoria, where sandy forelands have developed as the result of accretion of sand on either side of jetties built beside the entrance seventy-five years ago (Fig. 50). The pattern of accretion has been

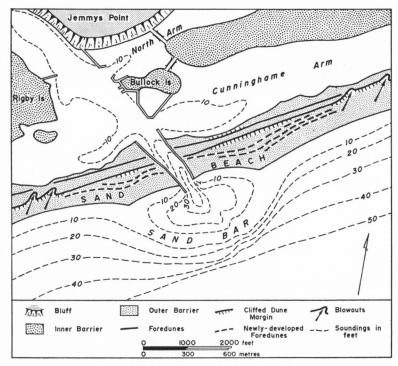

50 *Landforms at the entrance to the Gippsland Lakes, Australia (cf. Fig. 47), an artificial entrance cut in 1889 and maintained by current scour (maximum depth of entrance channel more than 20 m). Sandy forelands have developed on either side of the breakwaters that flank the entrance, and these show stages in parallel dune formation.*

determined by ocean swell, refracted on encountering a sand bar offshore, so that the shoreline of each foreland is cuspate in form. Three beach ridges have formed parallel to this prograding shoreline, and each has been colonised by dune grasses, which have contributed to the building of roughly symmetrical foredunes, each about 3 m high. The innermost now receives little sand, and here the dune grasses are being invaded by scrub.

The height and spacing of parallel dunes is a function of the rate of sand supply to the shore, the history of cut and fill and the effectiveness of vegetation in binding and building the dunes. Where sand supply has been rapid on a prograding shore subject to frequent storms, a large number of low, closely-spaced parallel dunes are formed, but on a similar shore less exposed to storms the effects of 'cut', which is responsible for the separation of the dunes into parallel ridges, are less frequent, and there are fewer, but larger, parallel dunes. In the absence of episodes of 'cut', a foredune becomes broadened as a coastal terrace of vegetated sand, advancing as the beach prograges. Where sand supply has been slow, parallel ridges are unlikely to form on a stormy shore, but a few low parallel dunes might be formed on more sheltered sections of the coast. Attempts have been made to deduce former sea levels from the heights of successive parallel dunes, an overall seaward descent of crests of swales being claimed as evidence of coastal emergence during their formation. Olson (1958) related alternating phases of backshore dune development and shoreline recession to historical fluctuations in the level of Lake Michigan. This kind of pattern could also be produced by diminishing sand supply or increasing storm frequency, without any change in the relative levels of land and sea, and so it cannot be accepted as conclusive evidence for coastal emergence. In recent decades there has been accentuated erosion on many Australian sandy shores, the seaward margins of many coastal dune systems being generally cliffed at the present time. This erosion may be the outcome of a phase of dominance of 'cut' over 'fill', either because of increasing storminess in coastal waters, or the progressive rise of mean sea level inferred from the evidence of tide gauges (Chapter II).

A succession of stable parallel dunes usually shows a sequence of vegetation types from grasses on the seaward side, through scrub to woodland or heath, accompanied by the development of successively deeper soil profiles. On quartzose dunes, pedogenesis (soil formation) leads to the development of podzols: the upper layers are leached of shell material by percolating rainwater which also removes the yellow stain of iron oxides from the sand grains, leaving the surface sand grey or white in colour, while the lower layers are enriched by the deposition of iron oxides, together with downwashed organic matter, to form a slightly-cemented red-brown sand horizon, often known as 'coffee rock'. The oldest parallel dunes on the landward side show the most advanced stages in vegetation succession, often with heath or heathy woodland on sand that has been deeply leached.

Many coastal dune systems do not show regular patterns of dunes parallel to the shoreline. Some show traces of a former parallel pattern that has been interrupted by the formation of blowouts and parabolic dunes (see p. 138), but others are quite

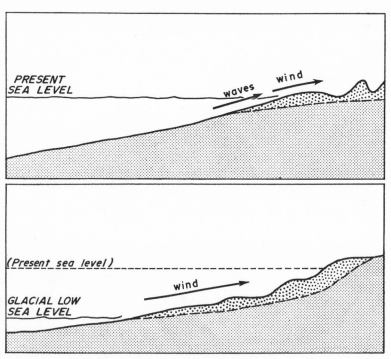

51 *Two possible mechanisms for the delivery of sediment to a coast. Above, with the sea at its present level, sand is being carried from the sea floor and delivered to the beach, where wind moves it inland; below, with sea level low during a glacial phase in Pleistocene times, sand is winnowed from the emerged sea floor and deposited in the coastal region.*

irregular, and have not necessarily originated from dunes built in parallel lines. There are good examples on the Atlantic coast of Europe and the Pacific coast of North America. The development of a large, irregular coastal dune topography can result from a rapid supply of sand to the shore, without the separation of deposited successive foredunes by cut and fill alternations in the manner described above, or from the delivery of sand blown from the emerged sea floor during glacial phases of the Pleistocene period, when sea level was lower (Fig. 51). This latter hypothesis may account for some of the dune topography off the S and W coasts of Australia, extending well below present sea level. Where aeolian calcarenite passes beneath present sea level, the dunes must have formed when the sea was lower, relative to the land, before a phase of coastal subsidence or sea level rise. The fringe of dune limestone plastered on to much of the coast of South Australia (Fig. 52) and Western Australia has been regarded as an accumulation of sand

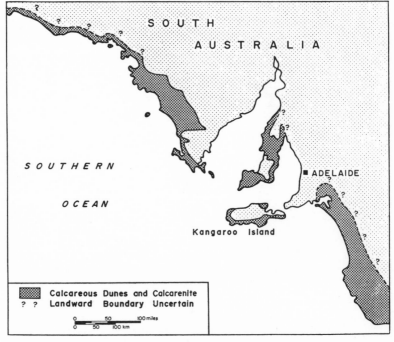

52   *Calcareous dunes and calcarenite on the coast of South Australia (after Bauer, 1961)*

blown from the emerged sea floor during Pleistocene phases of glacio-eustatically lowered sea level (Fairbridge and Teichert, 1952). It pre-dates the Holocene marine transgression, and a Pleistocene age is considered probable because of the presence of fossil remains of extinct marsupial species. It has usually been trimmed back on the present coast by marine denudation, forming rugged cliffs bordered by low tide shore platforms (Plate 18). Strongly-cemented layers, often associated with buried soil

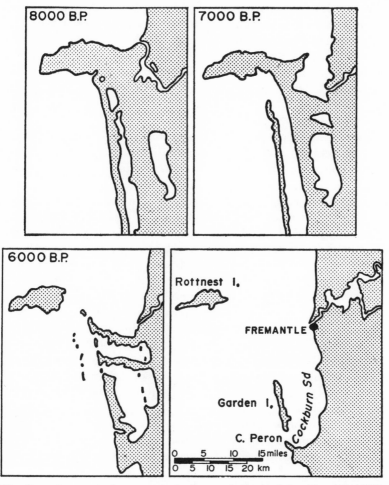

53 *Stages in coastal submergence in the Fremantle district (after Churchill, 1959)*

horizons, form prominent ledges in cliff profiles. Bauer (1961) has described how the calcarenite formed on Kangaroo Island, off the coast of South Australia, partly obscures the evidence of Pleistocene coastal terraces cut in bedrock.

The submergence and dissection of coastal dune limestone topography is well shown on the coast of Western Australia, where the Holocene marine transgression isolated Rottnest Island and Garden Island as it invaded Cockburn Sound and drowned the lower part of the Swan River valley (Fig. 53) (Churchill, 1959). The drowned ridges have generally been planed off by wave action on the sea floor, but detailed soundings have shown that parts of a submerged dune topography survive. Similar calcarenite topography is found on the S and E shores of the Mediterranean, on the Red Sea coast, around the Caribbean, on the Brazilian coast, and in South Africa.

Blowouts and parabolic dunes

Blowouts develop where the vegetation cover of unconsolidated coastal dunes is destroyed or removed, so that sand is no longer held in position. They are often initiated by intensive and localised human activity, where footpaths are worn over the dunes by people walking to a beach, or where trackways are made over the dunes for vehicles to reach the shore. Burning of coastal vegetation and excessive grazing by rabbits, sheep, cattle, or goats can destroy the vegetation cover and initiate blowouts, the effect of rabbits being accentuated where they have burrowed in the dunes. During a phase of aridity, weakening of the vegetation cover may lead to the formation of blowouts, and may mobilise dunes that had been fixed by vegetation under preceding humid conditions. Blowouts may also form when the outer margin of a foredune is cut away by the sea during a storm, leaving an unvegetated cliff of loose sand. On a prograding shore the rebuilding of the beach, with new berms developing into newer foredunes, prevents much erosion, but if the shoreline is generally receding, and the beach is not fully replaced, blowouts will continue to develop.

The evolution of blowouts is related to onshore winds and is most rapid on sections of the coast exposed to strong winds. A blowout that becomes enlarged begins to migrate through coastal dunes, with an advancing nose of loose sand (sloping at 30°–33°) and trailing arms of partly-fixed sparsely-vegetated sand; in this way it develops into a parabolic or U-dune (Plate 25). Dunes of this type are often found disrupting a pattern of parallel dunes (Fig. 54). In some areas they have been halted and fixed by vegetation in parabolic forms within an older system of parallel dunes. Active parabolic dunes, moving landward across a coastal dune fringe, are common on the SE coasts of Australia, and have been described in detail by Jennings (1957a) from King Island. Their movement is a response to the direction, frequency, and

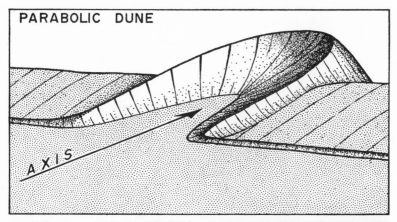

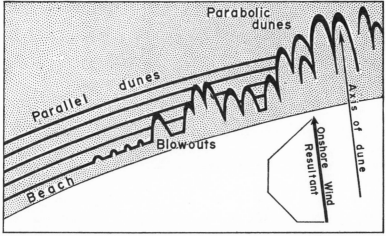

54 *Above, diagram of a parabolic dune; below, blowouts and parabolic dunes interrupting parallel dunes on a coast, and showing axes aligned with the onshore wind resultant (cf. Fig. 23)*

strength of onshore winds, the axis of each parabolic dune, defined as the line bisecting the angle between the trailing arms and directed towards the advancing nose (Fig. 54), being parallel to the resultant of onshore winds of Beaufort Scale 3 and over (Chapter V). This was demonstrated by Jennings on King Island where the parabolic dunes on the W coast are moving eastwards while those on the E coast are moving westwards. The axial directions of parabolic dunes behind the Ninety Mile Beach change as the orientation of the coast changes, bringing in different component groups of onshore winds with different angles of onshore resultant (Fig. 55)

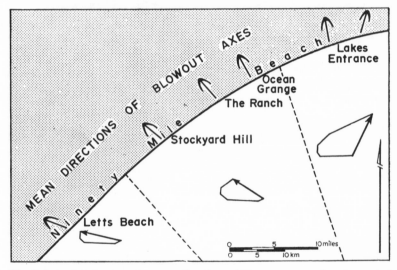

55   *Relationship between the mean directions of blowout axes and onshore wind resultants on three sectors of the Ninety Mile Beach, Victoria*

(Bird, 1965). Dunes retain a parabolic form as long as they remain partly vegetated, so that the trailing arms are held back by vegetation, but on parts of the Australian coast, overgrazing and fires have so reduced the dune vegetation that in places parabolic dunes have given place to masses of mobile, unvegetated sand, which show forms similar to dunes in arid regions, their profile at any time being a response to preceding wind conditions: a strong wind in one direction drives the crest of the dune one way, but subsequent winds from other directions modify the outline.

It is sometimes possible to distinguish successive waves of dunes which have migrated inland from a shore. On parts of the Australian coast, dunes have been banked against, or have climbed over, coastal cliffs, so that their crests reach altitudes hundreds of feet above sea level. Dunes have migrated up and over rocky promontories on the Victorian coast, notably Cape Bridgewater and Cape Paterson, and in some cases there are relict cliff-top dunes no longer linked with a source area of blown sand on adjacent beaches (Jennings, 1967). The very high dunes of the coast of S Queensland, on Stradbroke Island, Moreton Island (where at one point they exceed 275 m above sea level) and Fraser Island are not piled on to rocky foundations: above sea level they consist entirely of wind-blown sand, and borings have shown similar aeolian sand extending well below sea level. On Stradbroke Island there are several sets of transgressive frontal dunes, arranged in sequence parallel to the ocean coast, each partly overlapping its predecessor (Fig. 56) (Gardner, 1955). These have clearly advanced away from the shore during episodes of dune migration, but they are now stabilized beneath a cover of scrub and forest. On the New South Wales coast there are several sectors where a transgressive dune of mobile, unvegetated sand is migrating inland and burying older dune topography with its scrub or forest cover. Some of these dunes have been initiated by human activities, notably the grazing of stock and the burning of vegetation, liberating unconsolidated dune sand on the coastal margin. Residual hummocks, crowned by vegetation, are relics of the former stabilised dune topography. Some originated prior to European colonisation, for Captain Cook observed active dunes on Fraser Island in 1770. Aboriginal activities may have initiated these, or they may result from natural processes. A rapidly rising sea can trim back the margins of older dune topography, initiating transgressive dune development, but it is also possible that a rapidly receding sea could expose sandy deposits to be built into dunes and driven inland by onshore winds.

Similar sequences of transgressive dunes have been studied on the W coast of Auckland, New Zealand (Brothers, 1954), on the Pacific coasts of Oregon and Washington (Cooper, 1958), and on the Atlantic coasts of France and Britain. The Pyla dune, in SW France, is a huge mass of sand moving from the eroding shore S of the Arcachon estuary and spilling inland over the Landes pine forests. It was initiated during the eighteenth century, and has now

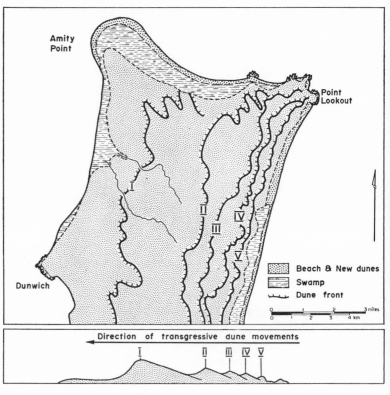

56 *Transgressive frontal dunes (I–V) on Stradbroke Island, Queensland. The diagrammatic cross-section shows their transgressive relationship (after Gardner, 1955).*

attained a crest altitude of more than 100 m, with a steep wall (32°) of sand advancing inland at an average rate of 1 m a year. It appears that tidal scour on the estuary shore, resulting from deflection of the ebb channel by the southward growth of Cap Ferret spit, maintains the sand supply to the migrating dune (Fig. 57). Newborough Warren, on the coast of Anglesey, consists of three transgressive dunes which have migrated inland from the sandy shore of Caernarvon Bay, and the Culbin Sands on the Scottish coast were formerly similar transgressive dunes until afforestation halted them.

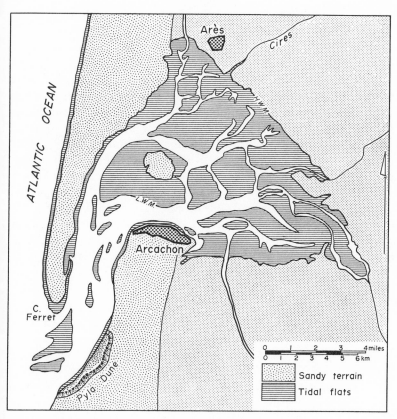

57　*The Bassin d'Arcachon and the Dune of Pyla, SW France*

Relative age of dune development

Some parts of the Australian coast show evidence of several phases of coastal dune accumulation, with older dunes on the landward side and newer dunes bordering the coast. Where the dunes are built of quartzose sand, there are marked contrasts between the topography, soils, and vegetation of older and newer dunes. Older dunes have a comparatively subdued topography, and have been leached of shells and iron oxides by percolating rainwater to a depth of several feet, underlain by indurated horizons of 'coffee rock', formed by deposition of downwashed iron oxides and organic matter. Their vegetation consists of heath, or heathy woodland. Newer dunes, forming a coastal fringe, are more continuous, with bolder outlines, and accretion of sand often still continuing. They are fixed by grasses or scrub, except where blowouts and parabolic dunes are developing, or where large transgressive masses of mobile sand are advancing inland. The sand is fresh and yellow in colour, and has not yet been leached of its small shell content or of the iron oxides which stain the sand grains. On the coast of King Island the older dunes on the landward side are thought to be of Pleistocene age, and the coastal fringe of newer dunes is the product of sand accretion during Recent times (Jennings, 1959). The junction between the two is well marked, particularly where the newer dunes are transgressive, advancing across the more subdued older dune topography. The distinction between older and newer dunes is sometimes demonstrable on calcareous dunes, where older dune forms are preserved in aeolian calcarenite, and newer dunes remain unconsolidated, but the contrasts in vegetation are less marked, and the older dunes have only a superficial layer of sand decalcified by percolating rainwater.

Coastal sandrock

Horizons of cemented sand, termed sandrock, occur in quartzose dunes on the E coast of Australia, and are sometimes exposed as ledges of slightly more resistant rock where the dunes have been dissected by wind action or eroded by the sea (Coaldrake, 1962). One type of sandrock is podzolic 'coffee rock', which is formed where iron oxides and organic matter have been washed down through the dune by percolating rainwater and deposited in a layer at, or slightly above, the water-table. Deposition of leached material at this level may be a consequence of repeated wetting and drying in the zone between the high-level water-table of the wet season and the low-level water-table of the dry season. Another type of sandrock has formed where sandy swamps or peaty sand in low-lying seasonally or permanently waterlogged sites have been overrun and compressed beneath advancing dunes. This type of sandrock may be distinguished on lithological grounds, for it often contains compressed plant remains, which are not found in 'coffee rock'.

Dune lakes

Hollows in dune topography that pass beneath the level of the water-table are occupied by dune lakes, which may be intermittent, forming only when the water-table rises after heavy rains and drying out subsequently. In seasonally dry areas the shrinkage of dune lakes may expose sand for deflation and construction of active and migrating dune forms moving away from the lake basin. Jennings (1957b) has described several types of dune lake from the coast of King Island including lakes impounded where dunes have been built across the mouth of a valley, ponding back the stream (Fig. 58A), and lakes occupying hollows excavated by deflation during dry weather (particularly between the trailing arms of parabolic dunes) (Fig. 58B). He found that dune lakes were most common on the older dune terrain, where the presence of underlying impermeable sandrock prevents the water draining away. Several

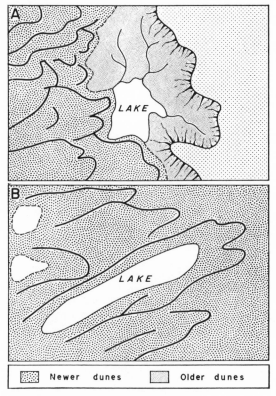

58 *Dune lakes formed by*
A *The impounding of a drainage system*
B *The flooding of a hollow formed previously by deflation*

lakes have been impounded in hollows in the older dune topography by the advance of transgressive newer dunes (Fig. 58C). Others originated as lagoons enclosed by coastal barriers when the sea stood at a higher level relative to the land (Fig. 58D).

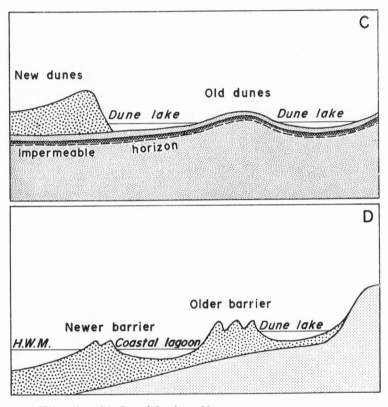

58 (continued) *Dune lakes formed by*
C *The migration of dunes across older dune topography*
D *The formation of barriers on a coast that has since emerged*
(After Jennings, 1957a)

VII

ESTUARIES AND LAGOONS

The marine transgression that took place during Recent times produced inlets at valley mouths and embayments on the sites of submerged coastal lowlands. The branched inlets formed by partial submergence of river valleys have been termed rias, and are well exemplified by Port Jackson (Sydney Harbour) and other inlets of similar configuration on the coast of New South Wales (Fig. 59).

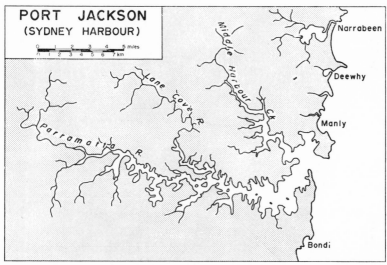

59   *Port Jackson, New South Wales, a valley system partly drowned by marine submergence*

As Cotton (1956) pointed out, these do not conform with Richthofen's original definition of a ria, which is restricted to drowned valleys aligned with a geological strike transverse to the coast, as in Bantry Bay and adjacent areas on the SW coast of Ireland. Even the Rias of NW Spain, from which the term originated, are not rias in this strict sense, and the term has come to be used as a synonym for drowned valley mouths remaining open to the sea: it overlaps broadly with the concept of an estuary (see below, p. 151). The drowned valley mouths which form branched inlets in SW England are rias in this wider sense. On the N coast of Cornwall they have been largely filled by sand washed in from the Atlantic Ocean, as at Padstow, but on the Channel coast they generally remain as relatively deep inlets, like Carrick Roads (Fig. 60), often with a threshold sand bar at the entrance, as at Salcombe.

Inlets at the mouths of formerly glaciated valleys on steep coasts are known as fiords, and are exemplified on the coasts of British Columbia and Alaska, Greenland, Norway, around the margins of the Arctic Ocean, and in W Siberia. The sea lochs of W Scotland

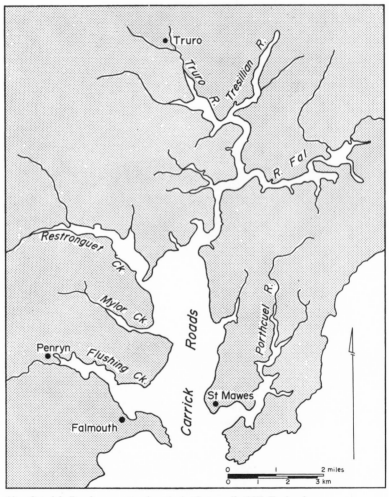

60   *Carrick Roads, an extensive ria in Cornwall, SW England*

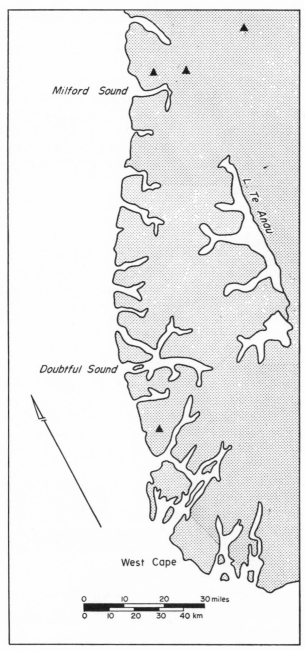

61   *The fiord coast of SW South Island, New Zealand*

are essentially fiords. In the southern hemisphere there are fiords on the S coast of Chile, on the W coast of South Island, New Zealand, where Milford Sound is a fine example (Fig. 61), and on the margins of Antarctica and outlying islands. The chief characteristics of fiords are inherited from prior glaciation. The glacial troughs were scoured out by ice action well below the low sea levels of Pleistocene glacial phases, and submerged by the sea as the ice melted away. Depths of more than 1300 m have been recorded in Scoresby Sound, a fiord on the E Greenland coast, and Sogne Fiord in Norway attains a depth of 1244 m. These inlets are steep-sided, narrow, and relatively straight compared with drowned river valleys, and they have the classic 'U-shaped' cross-profile of glaciated troughs, together with hanging tributary valleys which may also have been partly submerged by the sea. Near the seaward end there is often a shallower threshold, which may be either a rocky feature or a drowned morainic bar: the threshold at the entrance to Milford Sound is evidently a broad rocky sill. It has been suggested that thinner ice towards the mouths of these glaciated troughs scoured less deeply, leaving a threshold on the seaward side of an excavated basin. Some thresholds are less than 100 m deep, and were probably land areas separating the sea from a lake in the glaciated trough until submerged by the sea during the later stages of the Holocene marine transgression.

Inlets formed by marine submergence of valleys and depressions in low-lying glaciated rocky terrain are known as fiards in SW Sweden, and similar inlets in coastal plains of glacial deposition are termed förden on the Danish and German Baltic coasts. The firths of E Scotland appear to be a closely related category of drowned valley in formerly glaciated terrain. Beyond the limits of past glaciation, many drowned valley mouths have marginal features related to past periglaciation, as in SW England, where the rias are bordered by slopes covered with solifluction deposits (known locally as Head) exposed in fringing cliffs (see p. 57). The drowned mouths of deeply-incised river valleys beyond the limits of past glaciation may resemble fiords: Bathurst Channel, opening into Port Davey, in SW Tasmania (Fig. 62) was formerly thought to be a fiord, but it has been shown that this steep-sided inlet has no evidence of glaciation at or below present sea level, and none of the features of glacial sculpture that distinguish true fiords (Baker and Ahmad, 1959).

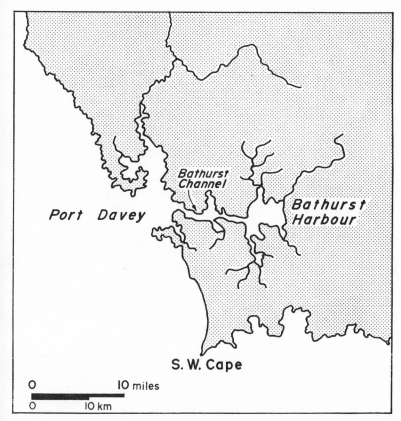

62 *The configuration of Port Davey, Tasmania. Bathurst Channel, formerly regarded as a fiord, is now considered to be a drowned river valley*

Estuaries

Estuaries are the mouths of rivers, widening as they enter the sea (Emery and Stevenson, 1957; Lauff, 1967). There is considerable overlap between the concept of an estuary and the various kinds of rias and fiords described above. Some rias and fiords are valleys that have been almost completely drowned by marine submergence, receiving so little drainage from the land that they are essentially arms of the sea, but most are fed by rivers, the mouths of which can be described as estuarine. Some have attempted to distinguish rias as features formed by the partial submergence of valleys incised into coastal uplands, restricting the term estuary to the mouths of valleys on low-lying coasts, but this is not always satisfactory:

Broken Bay, at the mouth of the Hawkesbury River in New South Wales, has many of the characteristics of an estuary, although it is steep-sided and bordered by high plateau country.

Alternatively, an estuary may be defined in terms of tidal conditions, as the lower reaches of a river subject to tidal fluctuations, or in terms of salinity, as the area where fresh river water meets and mixes with salt water from the sea. The ideal estuary is funnel-shaped, opening seawards, subject to tidal fluctuations, and influenced by the salinity of the sea (Fig. 63),* a form that is best

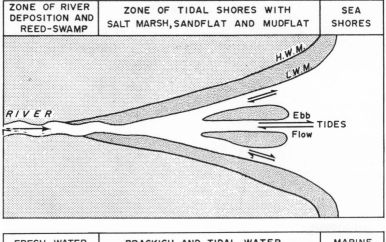

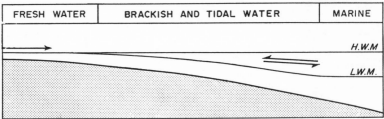

63 *Diagram of an estuary, showing zones related to salinity and tidal conditions*

developed on coasts with a relatively large tide range. In Britain the Thames, the Humber, and the Severn show this form; in W Europe the Rhine, the rivers of NW Germany, the Seine, the Loire, the Gironde, and the Tagus; and in Australia the rivers

* Sea salinity is generally in the range 33–38 parts of salt per thousand, mean ocean salinity being 35 parts per thousand.

draining into the macrotidal northern gulfs: the Fitzroy (King Sound), the Ord and the Victoria (Joseph Bonaparte Gulf), and the Daly.

At low tide in such estuaries, there are often a number of channels leading from the river to the sea, between banks of sand, silt, and clay. These are shaped largely by tidal currents. Some are predominantly ebb channels, with a residual outflow of water and sediment; others predominantly flood channels, with a residual flow into the estuary. The ebb channels become shallower towards the sea; the flood channels are deeper at the seaward end. Up to half an hour after low tide it is possible to find outflow continuing in ebb channels when the tide has begun to move up the flood channels (Robinson, 1960). The currents tend to change the position and dimensions of channels, cutting away sediment in one place and building it up in another, so that estuaries on which ports are sited often require dredging to maintain the depth and alignment of navigable approach channels. A knowledge of the characteristic patterns of erosion and deposition is necessary to ensure that dredged material is not dumped in such a place that it gets washed straight back into the channel required for navigation, and in recent years use has been made of radioactive tracers to determine paths of sediment flow in the Thames estuary to ensure access to the port of London. Banks of sand, silt, or clay exposed at low tide in estuaries are subject to rapid changes in configuration; they often bear superficial ripple patterns, produced by the action of waves and currents on the estuary floor as the tide rises and falls. Ahnert (1960) has drawn attention to the meanders that develop in estuarine channels, as in Chesapeake Bay, related to the patterns of strong ebb and flood currents at mid-tide level.

Estuaries are typically areas of active sedimentation, the area drowned by Recent submergence being progressively infilled, contracting in volume, depth, and surface area until the river winds to the sea through a depositional plain. It is possible to find every stage between the deep, branching estuary, little modified by sedimentation (e.g. Port Jackson), through to former estuaries that have been completely filled in, and with deltas developing beyond the former coastline (Chapter VIII). The chief factors are the original area and volume of the drowned valley mouth and the rate of sediment inflow from various sources. Enclosure by spits and barriers will tend to accelerate the natural reclamation of an estuary, whereas coastal subsidence will tend to postpone it.

Sediment is carried in by rivers, washed in from the sea, and derived from erosion of bordering slopes and shorelines. Rivers draining catchments where argillaceous formations outcrop, or where superficial weathering has yielded fine-grained material, are sources of estuarine muds. Many of the rivers of S and E England have derived muddy sediment from extensive clay lowlands for deposition in their estuaries: the London Clay has been a major source of Thames estuary muds, and the Mesozoic clay formations of the English Midlands have yielded muds to the estuaries of the Severn, the Humber, and the rivers draining into the Wash. The muds of the Wadden Sea are derived, at least in part, from the N German rivers, laden with silt and clay derived from loess and argillaceous outcrops. It is more difficult to account for the muds found in the rias of Devon and Cornwall, where the rivers drain mainly granite and Palaeozoic metasediment outcrops, but the source appears to be fine-grained material washed out of the Head deposits, the Pleistocene periglacial drifts which mantle hillsides in the SW peninsula, and are eroded as low cliffs bordering the rias. The same drifts mantle coastal slopes, but the fine-grained sediment is generally dispersed by the open sea, except in low wave energy sectors, as at Penzance and Newlyn on the W side of Mounts Bay, in the lee of the Lands End peninsula.

The rate of sediment yield from a river catchment is a function of lithology, weathering, and runoff conditions. It can be accelerated by deforestation, overgrazing, or unwise cultivation of erodible lands within the catchment, when stronger runoff augments the sediment supply to rivers, estuaries, and adjacent coasts. There are examples of this from many parts of the world. Rapid infilling of estuaries following soil erosion has influenced Chesapeake Bay on the E coast and several Oregon estuaries on the W coast of the United States. Soil erosion from deeply-weathered catchments in the humid tropics has delivered vast masses of mud to estuaries and coastal embayments, accelerating the progradation of mudflats and the spread of mangrove swamps: Djakarta Bay exemplifies this (Verstappen, 1953), and similar features are found on the W coast of New Caledonia, where estuaries are being rapidly infilled. On Mediterranean shores, shrinkage of estuaries since classical times is related to widespread soil erosion following the impoverishment of vegetation by overgrazing and unwise land use, notably in Greece. Mining operations can also accelerate sediment yield to rivers, as in Cornwall, where the wastes from tin, copper, and kaolin mining have been washed into the rivers and carried down to fill their estuaries (Everard, 1960). The River Fal is white with suspended kaolin derived from the St Austell claypits, which is being deposited in rapidly accreting marshlands at its entry to Carrick Roads (Fig. 60).

Deposition of marine sediment in the entrances to estuaries takes a variety of forms. The growth of spits and barriers across the mouths of estuaries has already been discussed (Chapter V). In open, funnel-shaped estuaries with a large tide range, marine sediment interdigitates with fluvial sediment in the zone traversed by ebb and flow channel systems as previously described. Where the entrance to the estuary is narrow, because of constriction by

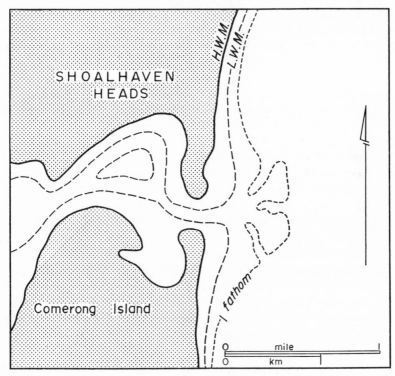

64   *The tidal delta at Shoalhaven Heads, New South Wales*

bedrock promontories or the growth of paired spits (p. 117), marine sediment is deposited in the form of a tidal delta, submerged at least at high tide, and having a pattern of diverging channels both on the landward and the seaward side. An example of a tidal delta has been described by Lucke (1934) from Barnegat Inlet on the coast of New Jersey, and a similar feature exists at the mouth of the Shoalhaven River on the New South Wales coast (Fig. 64). Where the tide range is small, and wave action strong, the tidal delta becomes a wider threshold, usually with a single channel winding over a broad sandy flat exposed at low tide, a smooth and gentle seaward slope and a steep inner slope, where sand is advancing into the estuary from the sea. Thresholds of this kind are found in the entrances to estuarine inlets on the New South Wales coast (Bird, 1967b), notably at Narooma (Fig. 68).

The configuration of an estuary can also be changed by wave and current action around its shores, particularly at high tide, when the sectors more exposed to strong wind and wave action tend to be eroded, and the material deposited on other parts of the shoreline, or on the estuary floor, in the manner described from Poole Harbour, an estuary in S England, by Bird and Ranwell (1964). Berthois has made detailed studies of erosion and sedimentation patterns at the mouth of the Loire (Guilcher, 1956), where much of the fluvial sediment is eventually deposited near high tide mark, after circulating within the estuary.

Salt marshes and mangrove swamps

Vegetation plays an important part in the shaping of depositional forms within estuaries and on tidal shores where estuarine conditions are found. Carpets of algae, such as *Enteromorpha*, and marine grasses, such as *Zostera*, partially stabilise the surface of tidal flats, and salt-tolerant (halophytic) plants colonise the margins of an estuary and spread forward across the inter-tidal zone as *salt marshes* (Chapman, 1960). These often show a well-defined zonation parallel to the shoreline, with plants such as *Spartina* or *Salicornia* dominating the outermost zone which is most frequently submerged by the tide, and the shore rush, *Juncus maritimus*, with other salt marsh communities, occupying higher zones that are less frequently submerged.

In tropical regions, and in many Australian and South African estuaries, mangroves form the pioneer community, spreading from

26 *Mangrove swamps on islands and bordering shores in an estuarine channel at Curtis Island, Queensland (Australian News and Information Bureau)*

27 Spartina townsendii *spreading on to mudflats in Poole Harbour, Dorset, England*

the shores of estuaries and colonising mudbanks exposed at low
tide (Plate 26). Drainage from the swamp becomes confined to
definite channels, often forming a reticulated creek network as in
the tidal estuarine sectors of the Nigerian coast (Allen, 1965),
or a dendritic pattern, as in many Australian mangrove swamps.
The extent to which mangroves promote accretion of mud is
uncertain, but where mud is being deposited mangroves spread on
to it and create a sheltered environment, enabling other salt marsh
plants to colonise when the mud surface has built up to a suitable
level. Mangrove encroachment is impeded by strong wave action
or tidal scour, so that mangrove swamps become most extensive

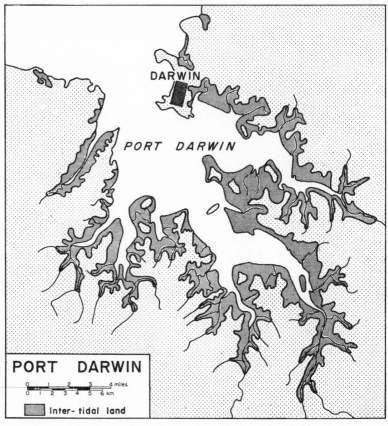

65   *Port Darwin, in N Australia, showing the extent of inter-tidal land in a gulf
where the spring tide range is large (5·5 m at Darwin)*

in sheltered areas, and only reach the sea on low wave energy sectors of the coast.

Sediment carried into salt marsh communities by the rising tide is filtered out by the vegetation and retained as the tide ebbs. In this way, the level of the land is gradually built up, and vegetated marshland encroaches on the estuary. Broad inter-tidal land has developed around the shores of gulfs and estuaries in N Australia, where the tide range is large (e.g. Port Darwin, Fig. 65), and wide tracts of mangrove swamp are backed by saline flats flooded during exceptionally high tides. Rates of accretion of sediment in tidal marshlands can be measured by laying down identifiable layers of coloured sand, coal dust, or similar material on the marsh surface, and returning later to put down borings and measure the thickness of sediment added. Measurements made in this way by Steers and others on salt marshes in Britain have shown that accretion is relatively slow at the upper and lower limits of a marsh, and more rapid in the intervening zone, where salt marsh vegetation forms a relatively dense sediment-trapping cover and is repeatedly invaded by sediment-laden tidal water (Steers, 1960). In section, salt marshes build up in the form of a wedge (Fig. 66), and, as

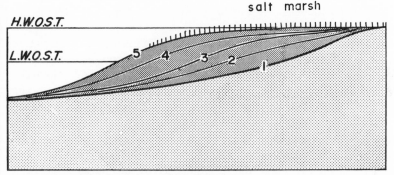

66   *Stages in the upward and outward growth of a salt marsh, showing levels of high and low water at ordinary spring tides*

they develop, the ebb tide maintains a system of creeks, by which water flows off the marsh (Plate 23). The margins of these may be built up as levees by accretion in a fringing shrub community, such as the *Halimione* which borders Norfolk salt marsh creeks. There may also be residual enclosed depressions, known as *pans*, which remain unvegetated and retain water after the tide has fallen. In arid regions high evaporation makes these hypersaline (more saline than the sea), or dries them out as salt flats. As the marshland surface is built upwards and outwards, a vegetation succession takes place, one salt marsh community giving place to another, until the land is built up to a level at which high tide flooding becomes rare. Sedimentation then proceeds very slowly, and if rain and runoff leach out the superficial salt, the marshland may be invaded by reed-swamp communities, dominated by such species as the common reed, *Phragmites communis*, and eventually by swamp scrub, or even woodland vegetation.

The outward spread of salt marsh into an estuary slows down as tidal waters become confined to a central channel, in which ebb and flow currents are relatively strong. The outer edge of the marshland may then oscillate, advancing during phases of local accretion, and receding during episodes of tidal scour, in relation to a dynamic equilibrium between channel dimensions and the volumes of tidal ebb and flow. Møller (1963) has reported detailed measurements of changes in marshland topography on the Danish coast, based on maps prepared in 1941 and 1959–62 on a scale of 1:10,000 with marshland surfaces contoured at 5 cm intervals. Patterns of vertical and lateral erosion and accretion can thus be located. Guilcher and Berthois (1957) measured sedimentation over a five-year period and recognised cyclic patterns of marginal erosion and accretion in the salt marshes bordering estuaries in Brittany. Continuing erosion of the seaward edge of a salt marsh can be due to several factors: current scour alongside a laterally migrating tidal channel, stronger wave action following deepening of the lower intertidal zone in consequence of progressive entrapment of sediment in the upper vegetated area, or continuing submergence of the coast. Blanc (1954) and Picard (1954) described the effects of vegetation on sedimentation in estuaries on the E coast of Corsica, where the local absence of salt marshes has been claimed as evidence of continuing subsidence of the land.

The influence of vegetation on estuarine sedimentation has been dramatically displayed in many estuaries in Britain since the arrival, or introduction, of the vigorous hybrid, *Spartina townsendii* (Plate 27), a grass which evidently originated as a cross between native and American species of *Spartina* in Southampton Water in the 1870s (Hubbard, 1965). It spreads across tidal flats of silt and clay, builds up marshland at a rapid rate, and has been used as a method of stabilising and reclaiming tidal flats in estuaries in various parts of the world, including the Netherlands, Denmark, estuaries in New Zealand, and the Tamar estuary in Tasmania. Older *Spartina* marshes show evidence of die-back, especially at their seaward margins, and the ecological reasons for this are not fully understood. As the sward dies away, sediment previously trapped is released and the marshland erodes away; the erosion is a sequel to die-back which must be due to some kind of adverse ecological condition. The process may be cyclic in the sense that released mud is deposited in new or reviving *Spartina* marshes elsewhere. Within Poole Harbour there are sectors where *Spartina* is still advancing, mainly in the upper estuary, as well as sectors of die-back and erosion of the marshland.

The tidal areas which lie behind barrier islands and spits on the N Norfolk coast and on the North Sea coasts of Holland, Germany, and Denmark have estuarine characteristics. There are banks and shoals of sand and mud, with marshland developed where vegetation has colonised the upper inter-tidal zone, and intricate and variable patterns of intersecting creeks out of which the ebb tide drains. Ripples (see also Chapter V) develop on the surface, especially in sandy areas, where they may attain a relief of more than 1 m usually aligned at right-angles to the tidal flow and migrating in the direction of the current (Van Straaten, 1953). Gierloff-Emden (1961) has published a series of aerial photographs illustrating features of the morphology of tidal shores in the Wadden Sea, on Germany's North Sea coast.

The mouths of many Australian estuaries are partly blocked by spits and barrier islands, built up by strong wave action on high wave energy coasts. Some are sealed off completely in the dry season, when river flow is unable to maintain an outlet, but after heavy rain the rivers flood, and water level builds up until it spills out, reopening the entrance (Jennings and Bird, 1967). Such estuaries are known as 'blind estuaries' in South Africa (Day, 1951). They are in fact estuarine lagoons.

Coastal lagoons

Lagoons partly or wholly enclosed by depositional barriers occur on many coasts. There are good examples on the Gulf and Atlantic coasts of North America, and on the coast of S Brazil; on the W African coast, notably at Abidjan and Lagos; the Landes coast in

France; in the Venice district, and on various parts of the Mediterranean (the Gulf of Lyons, the E coast of Corsica, and the Egyptian
coast), Black Sea and Caspian coasts; in NE Natal, the Coromandel
coast in India, and the W coast of Ceylon (Gierloff-Emden, 1961).
In Britain there are small lagoons behind shingle barriers at
Porthleven in Cornwall, Slapton in Devon, and Chesil Beach in
Dorset. On the Australian coast, lagoons are best developed behind
sandy barriers on the coast of Western Australia S of Perth, and
on the coast of SE Australia from the mouth of the Murray around
to S Queensland (Bird, 1967a).

Lagoons have formed where inlets or embayments produced
or revived by Holocene marine submergence have become enclosed by depositional barriers of sand or shingle. They show a
wide variety of geomorphological and ecological features, but their
essential characteristics can be summarised as in Fig. 67. There
are often three zones: a fresh-water zone close to the mouths of
rivers, a salt-water tidal zone close to the entrance, and an intervening transitional zone of brackish (moderately saline) but rela-

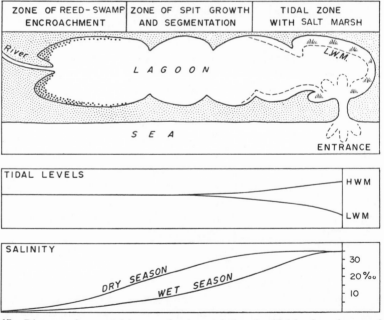

67 *Diagram of a coastal lagoon, showing variations in tidal levels and seasonal*
    salinity conditions

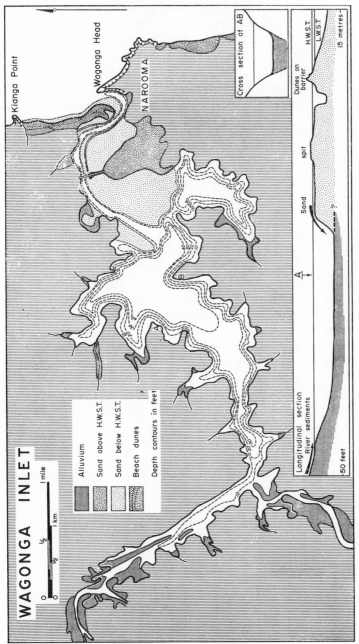

68 *Wagonga Inlet, New South Wales, an estuary with a well-developed sandy threshold*

tively tideless water. The proportions of each zone vary from one system to another: the Myall Lakes, on the New South Wales coast, consist largely of the fresh-water zone; Lake Illawarra consists largely of the intermediate zone, and Wagonga Inlet (Fig. 68) largely of the salt-water tidal zone. The extent of each zone depends largely on the proportion of fresh-water inflow to the lagoon system, lagoons tending to be more brackish in relatively arid regions, and on the nature and dimensions of the entrance from the sea. Lagoons completely cut off from the sea, like those on the Landes coast N and S of Arcachon, are essentially fresh-water lakes. In S England the Fleet is a typical estuarine lagoon, and Slapton Ley an almost fresh coastal lake (Fig. 28).

Lagoon entrances
The configuration of lagoon entrances is the outcome of a contest between the currents that flow through them and the effects of onshore and longshore drifting of sand or shingle which tend to seal them off (Bruun, 1967). Currents are generated through entrances in several ways. There are tidal currents produced by tides entering and leaving the lagoon, their strength increasing with tide range; there are currents due to outflow from rivers, particularly after heavy rain when floods build up the level of the lagoon so that water pours out through the entrance; and there are currents generated by wind action, onshore winds driving sea water into the lagoon and offshore winds driving lagoon water out to sea. Strong currents tend to maintain the dimensions of lagoon entrances. An entrance cut entirely in unconsolidated sediment has a cross-sectional area which varies in relation to the volume of water passing through, being widened or deepened during episodes of floodwater discharge. When the outflow is weak the entrance may be deflected by the longshore movement of sand caused by waves and wave-induced currents in the nearshore zone. Waves that arrive parallel to the coast move sand from the sea floor shorewards on to beaches and into lagoon entrances, and waves that reach the shore at an angle cause longshore drifting of sand, which deflects lagoon entrances and may seal them off altogether.

The position and dimensions of lagoon entrances change frequently in response to variations in the processes at work, and some have been stabilised by the construction of bordering breakwaters. When explorers arrived at the Gippsland Lakes, in Victoria, in the 1840s they found a small natural outlet at Cunninghame Arm, near the E end of the lagoon system, which was sealed in calm weather when onshore and longshore drifting of sand overcame the effects of transverse currents, and remained closed until heavy rains flooded the rivers and raised the level of the lagoons so that water spilled out over the barrier. The difficulties of navigating such an unreliable passage led to a local demand for an artificial entrance, and in 1889 a gap was cut through the enclosing barrier at what is now Lakes Entrance (Plate 28). This gap, bordered by stone jetties, has become a permanent artificial entrance, maintained by currents which have scoured a channel to a maximum depth of more than 20 m, and with a looped sand bar offshore testifying to the efforts of wave action to drive sand into the entrance and seal it off (Fig. 50). The natural entrance has fallen into disuse, and is now permanently sealed (Bird, 1965). In a similar way the previously migrating entrance to Ringkøbing Fiord, on the W coast of Denmark, has been stabilised by bordering breakwaters and a sluice at Hvidesande. Sand accumulating on the N side of the breakwaters is indicative of the southward drift which formerly diverted the lagoon entrance to the S.

Lagoon entrances are often located on sections of the shoreline where wave action is relatively weak, and the action of inflowing and outflowing currents therefore more effective. On wave refraction diagrams, such sections occur where there is a marked divergence of wave orthogonals, as at the head of the embayment in Fig. 6B. Many lagoon entrances on the New South Wales coast are situated at the S ends of sandy bays and close to rocky headlands, in positions where the dominant SE swell is intensely refracted, and therefore weakened (cf. Bascom, 1954). Other lagoon entrances are 'rock-defended'—close to offshore reefs or foreshore rock outcrops which break up constructive waves and prevent them completing the barrier at a particular point: the entrance to Lake Illawarra is protected by Windang Island, immediately offshore (Fig. 70).

The dimensions of lagoon entrances influence the extent to which tides invade a lagoon (tidal ventilation), the tide range diminishing rapidly inside the entrance. The more remote parts of large lagoon systems are unaffected by marine tides, but are subject to irregular changes of level from heavy rain, river flooding, or the action of strong winds. Winds blowing over a lagoon lower

28   *The artificial entrance to the Gippsland Lakes, Victoria, cut through the dune-covered coastal barrier (see Plate 30)*

29   *Reed-swamp encroachment on the shores of the Gippsland Lakes, Victoria. Phragmites and* Typha *are spreading into the lake, followed by swamp scrub vegetation on newly-formed sedimentary land.*

the level at the windward end and build it up to leeward; when the wind drops, the normal level is restored, often by way of a succession of oscillations of diminishing amplitude termed seiches.

Entrance dimensions also influence the pattern of salinity in a coastal lagoon. Salinity is determined by the meeting and mixing of fresh water from rain and rivers and salt water from the sea: it generally diminishes from the lagoon entrance towards the mouths of the rivers. In regions with seasonal variations of precipitation the salinity régime is also seasonal, for in the dry season sea water flows in through the entrance to compensate for the diminished river flow and the loss of fresh water by evaporation, but in the wet season evaporation losses are reduced and the lagoon is freshened by rain and rivers. In arid regions lagoons may become hypersaline, because the inflow of sea water is insufficient to prevent the development of high concentrations of salt as water evaporates from the lagoon. During the summer months, hypersaline conditions develop in enclosed lagoons such as Lake Eliza and Lake St Clair (Fig. 46), on the coast of South Australia, and also at the S end of the Coorong, away from the lagoon entrance, where tidal ventilation is too weak to prevent the development of high salinity in the evaporating water. Locally, saline evaporite deposits have formed where lagoons have dried out completely.

Salinity conditions influence modes of sedimentation. Clay carried in suspension in fresh water is flocculated and precipitated by the electrolytic effect of sodium chloride in solution when saline water is encountered. Salinity is also of ecological importance, affecting the development and distribution of shore vegetation around lagoons and thus influencing patterns of sedimentation in encroaching swamps.

Swamp encroachment

In the vicinity of tidal entrances to lagoons salinity conditions are essentially similar to those in estuaries. Banks of sediment are exposed at low tide, and the shores may be bordered by encroaching salt marshes and mangrove swamps. Away from the entrance, where the water is brackish and tidal fluctuations diminish, encroachment by salt marsh or mangrove swamp is much reduced; mangroves, in particular, are halophytic communities which require regular tidal inundation, and are best developed on the margins of tidal creeks (Gierloff-Emden, 1959). Lagoon shores in this inter-

mediate zone may be bordered by a narrow fringe of salt marsh
vegetation, but more often they are unvegetated and bordered by
beaches of sand or gravel derived from erosion of the shores.
Towards the mouths of rivers, where the water is relatively fresh,
reed-swamp communities dominated by species of *Phragmites*,
Scirpus, and *Typha* may colonise the shore, and where these communities are spreading into the lagoon, they influence the pattern
of sedimentation, trapping silt and floating debris and contributing
organic matter so that new land is built up.

   This process of swamp encroachment has greatly reduced the
area of relatively fresh lagoons, such as Lake Wellington, one of
the Gippsland Lakes (Fig. 47), which is bordered by extensive
swamp land on its S and W shores (Plate 29). Reed-swamp was
formerly more widespread on the shores of the Gippsland Lakes,
but increased salinity since the opening of the artificial entrance at

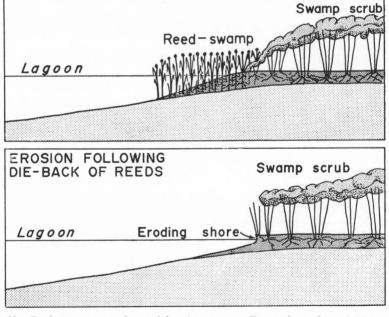

69   *Reed-swamp encroachment (above) promotes sedimentation and may initiate a
     vegetation succession to swamp scrub. If ecological conditions become unfa-
     vourable for reed growth (below), die-back occurs, and the swamp land that
     had previously been built up is likely to be eroded.*

Lakes Entrance has killed the reed vegetation and resulted in the erosion of land that had formerly been built up by sedimentation (Fig. 69) (Bird, 1965). On the other hand, freshening of lagoons that have become more completely sealed off from the sea, either naturally by the development of barriers or artificially by the insertion of weirs and barrages, can stimulate the spread of reed-swamp around their shores and initiate new patterns of sedimentation in the encroaching swamps. This has happened in the Murray-mouth lakes, following the completion in 1940 of barrages built to exclude sea water, and in the Étang de Vaccarès, a lagoon on the Rhône delta in the Camargue region of S France, where sluices now exclude sea water and freshening has followed the inflow of runoff from irrigated ricefields. Lymarev (1958) has described extensive *Phragmites* encroachment on the shores of the Aral Sea, drawing attention to the part played by swamp encroachment in promoting shoreline progradation there, and similar conditions are found on parts of the Baltic coast, notably in S Finland.

Configuration of lagoons
The initial form of a coastal lagoon depends on the shape of the inlet or embayment enclosed and the configuration of the barriers that enclose it. Some lagoons were originally broad embayments (Lake Illawarra, Fig. 70); others show the much-branched form of submerged valley systems (Lake Macquarie, Fig. 71); and in some the enclosing barriers incorporate islands (Tuggerah Lakes, Fig. 72). The inner shores of the enclosing barriers are usually simple in outline. There may be relics of recurved ridges, which marked stages in the prolongation of a spit which became a barrier enclosing an inlet or embayment; or there may be pro-montories formed where sand has blown over as an advancing dune, or washed over through low-lying sections of the barrier by storm waves or exceptionally high tides. Deposition is also common in the zone behind barrier islands where tides flowing in from en-trances meet, as in the Wadden Sea (Fig. 41) and on the S side of Scolt Head Island. The tidal watershed may become a marshland area.

As a barrier develops to enclose a lagoon, ocean swell is excluded and the estuarine effects of marine salinity and tidal changes of level are reduced. Winds blowing over the lagoon cause waves and currents which are related to the direction and strength and the

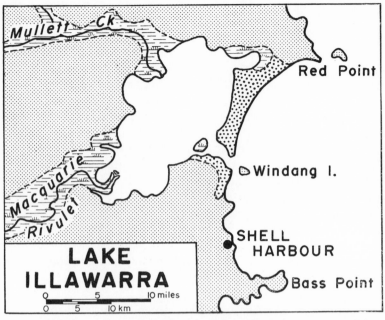

70 *Lake Illawarra, New South Wales, a broad embayment almost cut off from the sea by sandy coastal barriers, with a tidal entrance situated in the lee of Windang Island*

lengths of fetch across which these winds are effective. Long, narrow lagoons have the strongest wave action in diagonal directions, along the maximum fetch. If the shores are not protected by vegetation, waves coming in at an angle move sediment to and fro along the beaches, eroding embayments and building up spits, cusps, and cuspate forelands which may grow to such an extent that the lagoon becomes divided into a series of small, round, or oval lagoons, linked by narrow straits, or sealed off completely (Fig. 73). This process has been called segmentation by Price (1947), who described it from lagoons on the Texas coast, and Zenkovitch (1959) has described a similar process at work in some Russian coastal lagoons. Segmentation is essentially an adjustment of lagoon form to patterns more closely related to waves and currents generated within the lagoon. Currents play a part in smoothing the curved outlines of the shore in the later stages of segmentation and may also maintain the connecting straits be-

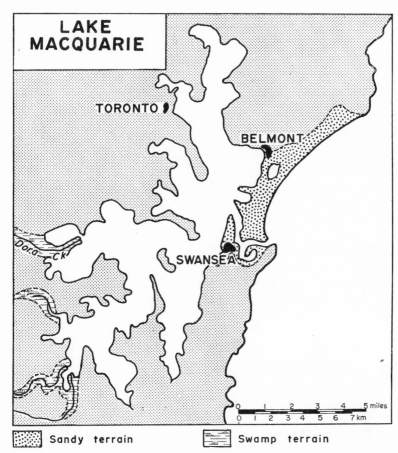

71 *Lake Macquarie, New South Wales, an embayment of intricate configuration almost enclosed by sandy coastal barriers*

tween segmented lagoons. But strong tidal currents deflect spit growth and inhibit the segmentation process.

In the Gippsland Lakes, Cunninghame Arm (Fig. 73) shows a series of cuspate spits (Plate 30) which, towards the E end, have grown to such an extent that they have almost cut off a chain of shallow pools, the Warm Holes. Small connecting creeks are maintained by wind-driven and tidal currents. On the South Australian coast, the lagoons between Robe and Beachport (Lake Eliza, Lake St Clair, and Lake George) (Fig. 46) have been formed by the segmentation of a long narrow lagoon, originally similar to the Coorong, farther north.

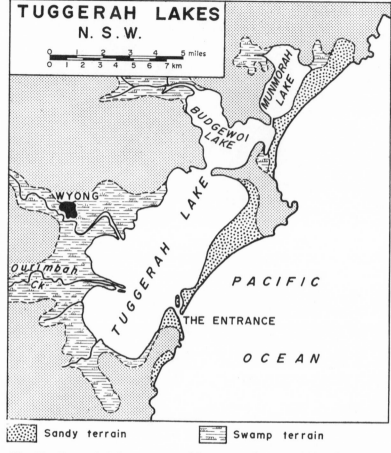

72 *The Tuggerah Lakes, a system of lagoons on the coast of New South Wales with an enclosing barrier which incorporates former offshore islands*

The sheltered waters of coastal lagoons are also favourable for the development of deltas at the mouths of rivers. In the Gippsland Lakes, small deltas have been built by the Latrobe and Avon rivers entering Lake Wellington, and the Tambo and Mitchell rivers entering Lake King. These and other deltas in coastal lagoons will be discussed further in the next chapter.

The patterns of sedimentation in lagoons are complex, and related to the characteristic patterns of waves and currents, as well as to the influence of shoreline vegetation. Comparison of the

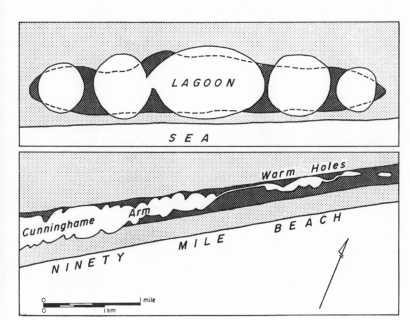

73 *Segmentation of a long narrow lagoon (above), the original shape of which is indicated by the pecked line (after Zenkovitch, 1959). The lower diagram shows segmentation in progress in the Cunninghame Arm, a branch of the Gippsland Lakes, Victoria (cf. Fig. 47).*

geomorphological features of lagoons on the Australian coast indicates that those bordered by encroaching reed-swamp contract in area until they are completely occupied by swamp land, whereas those without shoreline vegetation are re-shaped by waves and currents as segmentation proceeds. Eventually, after many changes in configuration, coastal lagoons receiving sediment are replaced by coastal plains, across which rivers and residual creeks wind, uniting to pass out to sea through the tidal entrance. Close examination of patterns of sediment on coastal plains sometimes reveals the outlines of former lagoon systems that have been extinguished by deposition of sediment. Alternatively, renewed erosion may remove the barrier, opening up the lagoon as a coastal embayment. It is likely that Guichen Bay, at Robe, on the South Australian coast (Fig. 46) and Rivoli Bay, at Beachport, farther south, were at one stage enclosed lagoons, comparable with Lake Eliza, Lake St Clair, and Lake George, which lie behind a barrier of calcarenite on the intervening coast. Reefs and islands of calcarenite in Guichen Bay

30 *The coastal barrier near Lakes Entrance, Victoria, showing cuspate sandspits on the shore of Cunninghame Arm (above), scrub vegetation on the fore-dune, and the Ninety Mile Beach (below)*

31 *The delta built by Macquarie Rivulet into Lake Illawarra, New South Wales*

32 *Beach rock exposed on the shore of a sand cay at Low Isles, off the Queensland N coast*

and Rivoli Bay indicate the former extension of this barrier north-ward and southward, and it is inferred that these sections have been removed by marine erosion. Fig. 53 traces the breaching of similar calcarenite barriers on the coast of Western Australia to form Cockburn Sound, and lunate embayments of similar origin have also developed on the calcarenite coast of Israel near Nahsolim, and on the N coast of Puerto Rico (Kaye, 1959).

Detailed studies of the geomorphology and ecology of coastal lagoons include Webb's (1958) work on Lagos lagoon, Stevenson and Emery (1958) on Newport Bay, California, and Fisk (1958) on the Laguna Madre, a hypersaline Texas coast lagoon. In terms of the geological time scale, coastal lagoons are ephemeral features, likely to be replaced by depositional plains or reopened as coastal embayments, depending on the subsequent evolution of the coastal region in which they have developed.

VIII
DELTAS

Deltaic forms and structures

Deltas are built where sediment brought down by rivers has filled in the mouths of valleys drowned by Holocene marine submergence to form a depositional feature protruding from the general outline of the coast (Samojlov, 1956). A delta will form if the sediment accumulating at the river mouth exceeds that carried away by waves and currents. Most deltas also incorporate sediment of marine origin, brought alongshore or from the sea floor. Small rivers may build deltas on the sheltered shores of lakes or tideless seas, but on coasts dominated by strong wave action and tidal scour, protruding deltas are built only by large rivers draining catchments that yield an abundance of sediment. Sectors of formerly submerged valley mouths reclaimed by fluvial deposition are essentially deltaic, even if they do not protrude from the general outline of the coast. The Rhine, Guadalquivir, Senegal, Colorado, and Amazon all have features of this kind, and it is convenient to include them as deltas, even though they lack the characteristic shape that led Herodotus to apply the Greek letter as a geomorphological term.

The classic example of a delta to which Herodotus referred is that of the Nile, built into the Mediterranean Sea on a coast where the tide range is less than 1 m. Other examples of protruding deltas built into relatively tideless seas include the Ebro, the Rhône, and the Tiber deltas built into the Mediterranean, the Po delta in the Adriatic Sea, the Danube delta in the Black Sea, and the Volga delta in the Caspian. Microtidal conditions are not essential for delta development, however: the Irrawaddy delta region has a tide range of 5·5 m and the Ganges 4·5 m. Both are on coasts where the wave energy is relatively low, because of broad shallow areas offshore which diminish wave action. The Mississippi delta has been built in a microtidal and generally low wave energy environment.

Deltas are rare on high wave energy coasts, where sediment delivered by rivers is dispersed by strong wave action, the coarse fraction (sand and gravel) being deposited in beach and barrier formations on the coast, and the finer material settling on the sea floor or carried away in suspension by the sea. Protruding deltas have generally not developed on the ocean coasts of Australia because of high wave energy conditions and because fluvial discharge is relatively small: an exception is the De Grey delta, on a sector of the NW coast with moderate wave energy. Many of the rivers have a marked seasonal régime, with reduced flow and reduced sediment yield in the dry season, and several valley mouths drowned by Holocene submergence have not yet been filled in by sediment. The Murray-Darling system has a catchment basin of more than one million sq.km, but in its lower reaches the Murray flows through a semi-arid region and loses so much water by evaporation that it has been unable to carry sufficient sediment to fill in the lagoons (Lakes Albert and Alexandrina) that remain at its mouth. In SE Australia many submerged valley mouths remain as inlets or lagoons enclosed by spits or barriers, but larger rivers

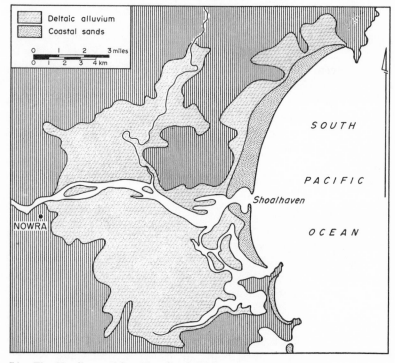

74   *The Shoalhaven delta, New South Wales*

such as the Snowy and the Shoalhaven have brought down more sediment, and reclaimed their drowned valleys as depositional plains. These deltaic plains do not protrude from the general alignment of the coast; their seaward shores have smoothly-curved outlines shaped by ocean swell (Fig. 74). On low wave energy sectors of the Queensland coast, sheltered from ocean swell by the Great Barrier Reefs, rivers that have reclaimed their drowned mouths are building protruding deltas (e.g. Burdekin, Fig. 80).

The shape of a delta is determined partly by sediment supply and partly by the action of waves and currents (Gulliver, 1899). Where the supply of fluvial sediment is abundant, and wave and current scour have not been excessive, *lobate* deltas are formed (e.g. the Nile, Fig. 75). A rapid supply of sediment on a low wave energy coast, as in the Mississippi delta (Russell and Russell, 1939), leads to the formation of *digitate* deltas where narrow levees are prolonged at the mouths of rivers during floods. Increased wave and current scour leads to the formation of *cuspate* rather than lobate deltas, the flanks presenting concave rather than convex

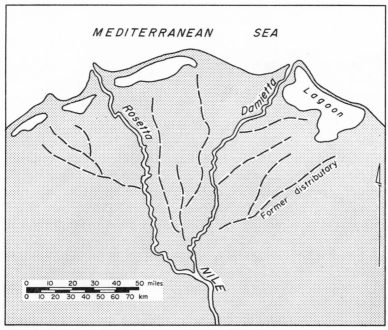

75   *The Nile delta, Egypt*

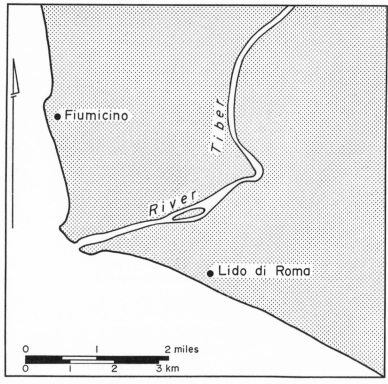

76   *The Tiber delta, Italy*

outlines (e.g. the Tiber, Fig. 76). Some deltas consist of a single river mouth flanked by depositional land while others have wandering and intermittent outlets, as in the Rhône delta, where the old main outlet is now the relatively unimportant Petit Rhône and the present river discharges through the larger Grand Rhône channel farther east (Fig. 77). Others have a river branching into several distributaries, as in the Nile delta, where all but two of the outflow channels are now defunct (Fig. 75). Distributaries are common on deltas where the accumulation of sediment in river channels tends to lift the river to the surface, so that it spills out in various directions. Some distributaries originate where new outlets are scoured during floods; others where a river mouth is blocked by deposition or split into two or more channels by the formation of shoals and islands, as at the mouth of the Fly, flowing into the

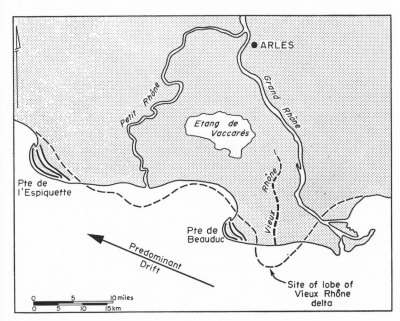

77   *The Rhône delta, France*

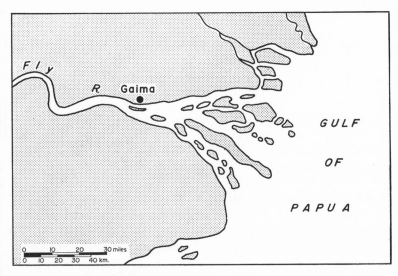

78   *The Fly delta, New Guinea*

Gulf of Papua (Fig. 78). Tidal ebb and flow tends to maintain
distributary mouths, whereas longshore drifting tends to divert
them or seal them off altogether. River outlets may change during
floods, as in the Rio Sinu on the Colombian coast, which breached
a levee during a 1942 flood and has since built a new delta into an
adjacent embayment (Troll and Schmidt, 1965). The Mississippi
has had a long history of distributary development, with successive
formation of subdelta lobes. As a lobe extends, channel gradients
in the river are reduced, and the outflow diminishes. Bordering
levees may then be breached during floods to form crevasses,
giving a shorter and steeper outflow channel, which then develops
a new distributary lobe, producing eventually the branched form
shown in Fig. 79.

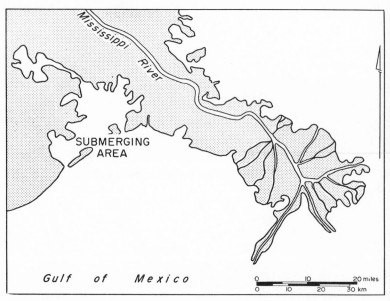

79   *The Mississippi delta, U.S.A.*

Beaches, spits, and barriers built by marine processes on the
shores of a delta often enclose lagoons and swamps, and are sometimes themselves incorporated as the delta grows larger. The
Burdekin River has built a large protruding delta (Fig. 80) filling
a former embayment and enclosing a number of high islands that
formerly lay offshore. The river divides into distributaries in the

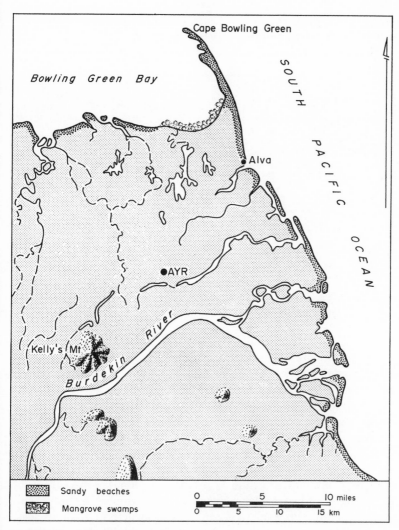

Bowling Green Bay

Cape Bowling Green

SOUTH

PACIFIC

OCEAN

Alva

AYR

Kelly's Mt

Burdekin River

Sandy beaches

Mangrove swamps

| 0 | | 5 | | 10 miles |
| 0 | 5 | 10 | | 15 km |

80   *The Burdekin delta, Queensland*

delta region, some of which are former outlets, partly sealed, or carrying water only during floods. A number of spits have been built northwards, deflecting outlets from former distributaries, and the growth of a long recurved spit, terminating in Cape Bowling Green, also testifies to a northward longshore drift resulting from the prevalence of waves generated by the SE trade winds. In N

Italy the Tagliamento delta has a series of symmetrical parallel
sandy beach ridges marking stages in the progradation of a cuspate
delta shoreline (Fig. 81). The cheniers of the Mississippi delta are

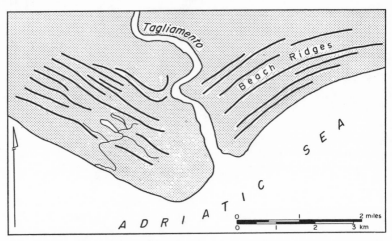

81   *The Tagliamento delta, Italy*

of different origin, being long, narrow, low-lying strips of sand
deposited on pre-existing swamps and alluvial flats, and marked
out by contrasts in the vegetation pattern (Russell and Howe,
1935). Similar features are found in Surinam on the N coast of
South America, where several sand ridges have been emplaced on
broad swampy deltaic plains adjacent to the Amazon and the
Orinoco. It is probable that these were emplaced during occasional
high tides or storm surges when sand was carried inland on coasts
where wave energy is normally low (Price, 1955).

Delta shorelines continue to prograde as long as the supply of
sediment, mainly from the river, exceeds removal by wave and
current action. Progradation can be spectacular in shallow seas,
the Volga delta prograding 170 m annually, but if deep water, or
a submarine canyon, exists offshore, as off the Ganges delta, sediment is lost to the sea floor and delta growth may be relatively
slow. Sectors of a delta shoreline no longer receiving sediment may
start to be eroded, as on the Rhône delta, following the decay of
the Vieux Rhône and diminishing outflow from the Petit Rhône
(Fig. 77). Lobes of sediment which had been built by these distributaries have been consumed by marine erosion, and the

dominant SE waves have transported derived sandy material westwards to be built into depositional forelands at Pointe de Beauduc
and Pointe de l'Espiguette respectively. Changes in delta configuration can sometimes be traced from old maps, as shown by
Nossin (1965a), who used cartographic records to reconstruct the
evolution of the North Padang delta in Malaya since the early
seventeenth century.

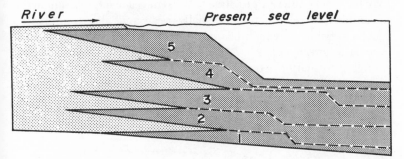

82   *The stratigraphy of a delta that has been subject to five successive marine
transgressions, each marked by an intercalation of marine sediment. Renewed
growth of the delta after each transgression has produced a wedge of fluvial
sediment.*

The simplest deltas are those built by rivers entering tideless
lakes. Where the lake level has fallen, and bordering deltas have
emerged, it is possible to examine their sedimentary structure.
The emerged delta that bordered the ancient 'Lake Bonneville'*
consists of almost horizontal bottomset beds, which were laid down
on the lake floor, inclined foreset beds (10°–25°), consisting of
layers of sediment that were delivered progressively as the delta
advanced, and horizontal topset beds, which form the upper
surface (Gilbert, 1890). Deltas built up on sea coasts during and
since Holocene marine submergence are more complicated than
this and may not include foreset beds. Their sedimentary structures, inferred from borings, show intercalations of marine sediment
deposited during episodes of marine transgression (Fig. 82). Such
a delta has been built forward during phases of stillstand of sea
level (or during episodes when sea level was falling), but at other

* Lake Bonneville is the name given to an extensive lake that existed in W Utah
(U.S.A.) during Pleistocene times. This region has since become more arid, and
the lake is now represented by scattered remnants, of which Great Salt Lake is
the largest.

times it has been submerged by a rising sea. Much of our knowledge of the oscillating transgression of the sea in late Pleistocene
and Recent times has been obtained from the stratigraphic evidence
yielded by borings in deltas and coastal plains formed by deposition,
notably on the Gulf Coast of the United States and in the Rhine
delta region. In dealing with this kind of evidence, it is necessary
to remember that the level of stratigraphic horizons within a delta
may have been lowered, relative to present sea level, by compaction
of underlying sediments (especially peat and clay) and by the
crustal subsidence which has taken place in many deltaic regions.
The stratigraphy of the Mississippi delta, for example, shows
plunging formations equivalent to deposits on river terraces which
border the valley upstream (Fig. 83), and sectors of this delta

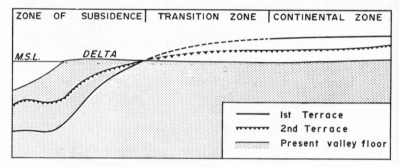

83   *Plunging terraces traced into a delta, where borings have indicated deposits
      equivalent to those on the 1st and 2nd terraces upstream (based on Russell's
      interpretation of the Mississippi delta)*

where continuing subsidence is not compensated by sedimentation become lagoon depressions, or are invaded by the sea. On the
Niger delta a sequence of dead coralline banks, partly obscured
by sediment on the sea floor, indicate stages in subsidence and
lateral tilting during the past four thousand years (Allen and Wells,
1962). The volume of sediment deposited in these deltas in Recent
times is enormous: 2800 km³ in the Mississippi delta, and 1400
km³ in the delta of the Niger.

   The valley floors built up by deposition of sediment in drowned
valley mouths around the Australian coast show deltaic stratigraphy,
with interdigitation of fluvial, estuarine, and marine sediments.
The sedimentary fill normally occupies a valley that was excavated
by the river when it flowed out to lower sea levels during glacial

phases of the Pleistocene period, and may include relics of materials deposited in these valleys during interglacial phases of high sea level, as well as the sediment deposited during and since the Holocene sea level rise.

It is not easy to discover how far the sea penetrated into river valleys during the Holocene submergence. Stratigraphic evidence, where available, is the most reliable indication, the extent of marine deposits beneath the valley floor marking the maximum extent of submergence. These deposits are overlain by fluvial sediments deposited by the river since the submergence came to an end. Often horizons of marine deposits pass laterally into estuarine, then fluvial deposits, and the limit reached by the sea can rarely be determined accurately. Where the valley sides have been cliffed, and are bordered by beach deposits containing marine fossils, it is evident that open water existed for a sufficient period for waves to attack the margins of a drowned valley, before river deposition built up the flood plains and deltas. In some valleys the maximum extent of submergence may be indicated by the form of the meandering river channel: upstream, where the valley floor was never submerged by the sea, meandering is often more intricate than in the lower reaches, where broad, sweeping channels wind across a valley floor that is strictly of deltaic origin.

Natural levees

Reference has been made to the existence of natural levees bordering river channels and backed by low-lying, often swampy or flooded depressions in the lower parts of river valleys. This topography is the outcome of unequal building up (aggradation) of the flood plain, whereby sediment is deposited more rapidly in the zone immediately adjacent to the river channel than in the zone beyond. When a river rises and overflows its banks, the flow of water is most rapid along the line of the river channel and much slower on either side so that the coarser load of sand and silt carried by flood-waters is relinquished at the borders of the channel where water velocity diminishes, and only the finer clay particles are carried into the calmer water beyond. As natural levees are built up along the sides of the river channel, they develop gentle outward slopes, passing down to lateral depressions, known as levee-flank, back-plain, or backswamp depressions (Russell and Russell, 1939). After floods have occurred, water may persist in these depressions for

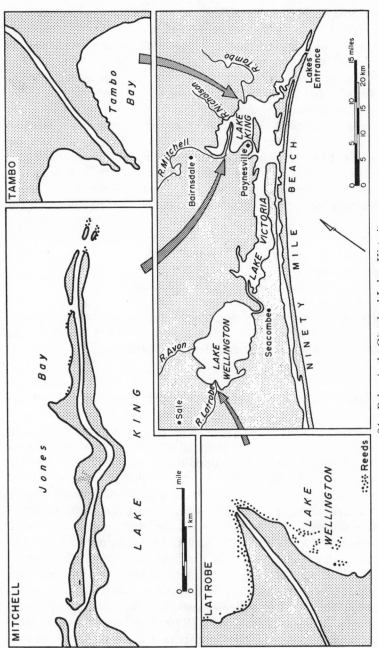

84 *Deltas in the Gippsland Lakes, Victoria*

a long time, particularly if the levees impinge locally on the valley side and enclose them, completely preventing down-valley drainage. The depressions are floored by clay deposits which settle from flood-waters, and are often occupied by swamp vegetation, which may build up peat deposits. Similar valley-floor features are found in SE Australia, where the rivers have régimes with marked seasonal variation and flooding occurs almost every year. In dry years, repeated evaporation of water from enclosed backswamp depressions leads to a concentration of salt, increasing soil salinity so that some of these backswamp depressions are occupied by salt marshes, or even unvegetated saline flats. Deflation from these dry depressions may lead to the building of silt or clay dunes, as on the Senegal delta (Tricart, 1955), and seasonally dry distributaries yield sand for local dune development.

The process of deposition which produces natural levees along the sides of a river channel also prolongs them at the mouth of the river, providing wave and current scour does not prevent this. In New South Wales, sedimentary jetties have formed at the mouths of creeks opening into coastal lagoons. The delta of Macquarie Rivulet, in Lake Illawarra (Fig. 70), is curved and elongated, with a branched outlet (Plate 31), and similar features have developed at the mouths of Dora Creek in Lake Macquarie (Fig. 71) and Ourimbah Creek in the Tuggerah Lakes (Fig. 72). In the Gippsland Lakes (Fig. 84) the Latrobe delta is lobate, the Tambo delta cuspate, and the Mitchell delta consists of long silt jetties. The growth of these deltas has been assisted by the presence of shoreline reed-swamp, which traps and retains sediment deposited at the mouths of rivers during floods. Where the reed-swamp fringe has disappeared as the result of salinity increase, the deltaic sediment is no longer retained by shoreline vegetation, and waves are attacking the unprotected shores of the silt jetties, eroding their margins and dissecting them into a chain of islands. The Tambo delta and the Mitchell delta in Lake King show advanced stages in dissection, but the Latrobe and Avon deltas, in Lake Wellington, are still reed-fringed and still growing. Evidently the presence of reed-swamp influenced the pattern of sedimentation so that deltas were built at these river mouths, but when the reed fringe disappeared, the deltas were no longer stable, and erosion began to consume them (Bird, 1965). This is a special case. In general, abundance of sediment and absence of excessive wave and current scour are sufficient to ensure the formation of a delta.

IX

CORAL REEFS AND ATOLLS

Reefs built by coral and associated organisms are characteristic of tropical waters, and are widespread between latitudes 30°N and 30°S in the W parts of the Pacific, Indian, and Atlantic oceans. They are well developed in the Caribbean Sea, the Red Sea, and Indonesia. In this account, examples are drawn mainly from the reefs and atolls of the SW Pacific area, and around the coasts of Australia (Fig. 85), particularly off the E coast of Queensland where the Great Barrier Reefs extend from Torres Strait in the N to the Bunker and Capricorn groups in the S (latitude 24°S). Outlying reefs extend farther south, to the Middleton and Elizabeth atolls (29°–30°S) and the W coast of Lord Howe Island (32°S). There are scattered reefs off the N coast of Australia, and off the W coast of the continent they extend as far south as Houtmans Abrolhos, in latitude 29°S (Fairbridge, 1967). Where reefs border the coast they are termed fringing reefs; where they lie offshore, enclosing a lagoon, they are known as barrier reefs; and where they encircle a lagoon, they are called atolls (Fig. 86). In addition there are isolated reef platforms of various shapes and sizes which do not fit any of these categories.

Reefs are mainly of biological origin, built by coral polyps, small marine organisms that take up calcium carbonate from sea water and grow into a variety of skeletal structures. Closely associated with these are small plants, the calcareous algae, which grow in and around the coral structures. Each set of organisms assists the other, the algae drawing nutrients from coral, and utilising much of the carbon dioxide released into the sea by coral respiration in the manufacture of food by plant photosynthesis; this in turn replenishes the oxygen dissolved in sea water, and thus maintains a supply of oxygen for coral respiration. While coral is essential for reef-building, it generally forms only a small proportion of reef material. Algae, together with foraminifera, molluscs, other shelly organisms, and sedimentary fragments of shell and coral, fill in the skeletal coral structures to form a calcareous mass, which eventually, with partial solution and re-precipitation of calcium carbonate, becomes a massive reef limestone. There are

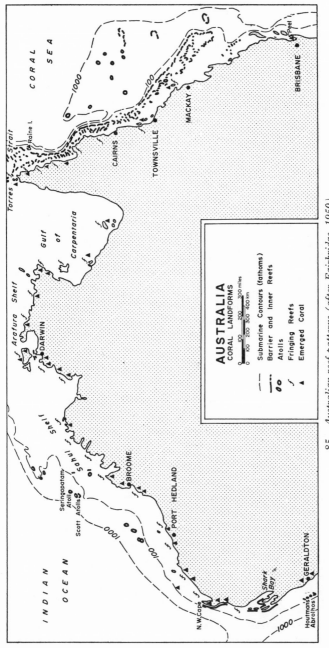

85 *Australian reef patterns (after Fairbridge, 1950)*

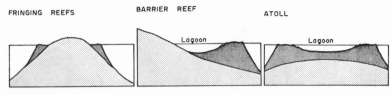

86   *Reef landform types*

also extensive sediments derived from reef erosion banked against
the flanks of the massive reef.

Factors influencing reef development

Reef-building takes place in ocean areas where free-floating plank-
tonic coral larvae are distributed by ocean currents and where
ecological conditions permit the establishment and growth of
coral and associated algae. An adequate supply of sunlight is
essential for algal photosynthesis, and growth of coral is best in
clear, warm water. Intensity of sunlight diminishes downwards
into the sea, and although live corals have been found in exceptionally clear water at depths as great as 100 m, the maximum
depth at which reefs are being built is rarely more than 50 m.
In coastal waters, increased cloudiness due to land-derived sediment (chiefly silt and clay) in suspension reduces penetration by
sunlight and impedes growth of reef-building organisms. Off river
mouths, this cloudiness, together with the blanketing effects of
deposits of inorganic sediment, often prevents the growth of reef-
building organisms and breaks the continuity of fringing reefs.
Corals can dispose of small accessions of sediment, but are choked
by continued deposition of large quantities of detritus, or by heavy
loads of sediment dumped suddenly from discharging flood waters.

Coral growth is confined to warm seas, where the mean temperature of the coldest month (generally August in Australia) does
not fall below 18°C. Salinity must be within the range 27–38 parts
per thousand, dilution by fresh water off river mouths contributing
to the persistence of gaps in bordering reefs. Excessive salinity may
explain why reefs are absent from certain parts of the coast; the
absence of coral in Hamelin Pool, in Shark Bay (Western Australia),
was attributed by Fairbridge (1950) to excessive salinity, up to
48 parts per thousand in summer. Where ecological conditions are
favourable and there is an adequate supply of mineral nutrients in
the sea water, reefs grow upwards until they form platforms that

are exposed at low tide, but the corals cannot survive prolonged exposure to the atmosphere and they die off when the reef platform attains a level a foot or so above low water spring tide level. Algae (chiefly *Lithothamnion*) may then build up a slightly higher reef crest, awash at high tide.

Coral growth remains vigorous on the steep bordering slopes of reefs below low tide level. The best growth is where sea water is being circulated sufficiently to renew the supply of plankton, on which the coral polyps feed, and to maintain the supply of oxygen, particularly at night, when it is no longer replenished by algal photosynthesis. Where the slopes of reefs are exposed to strong wave action, as on the seaward flank of a barrier reef or the outward slope of an atoll, there is adequate circulation of sea water for vigorous coral growth, but the impact of waves tends to break off, or inhibit the formation of, the more intricate skeletal forms of coral, so that the seaward slopes often consist of a more compact mass of growing coral. On the leeward side, in more sheltered water, a greater variety of growth forms is found, and some of these, broken off during occasional storms, are cast up on to the reef platform as sand and gravel fragments, and occasionally the large round boulders that Matthew Flinders termed 'negro-heads'. Large quantities of broken reef material are banked up as sedimentary deposits on reef flanks, and on lagoon floors, and some of this material is deposited as a veneer on the platform exposed at low tide, and may be built up into islands surmounting the reef and extending above high tide level. Off the Queensland coast these are known as 'low islands' to distinguish them from the 'high islands' of non-coralline rock that are also present (Plate 21). 'Low islands' will be discussed later.

Fringing reefs

The simplest reef landforms are fringing reefs, built upwards and outwards in the shallow seas that border continental or island shores. They consist of a platform at low tide level which is similar in many ways to the low tide shore platforms found on cliffed limestone coasts (Chapter IV), although it has been built up, rather than cut down, to this level. Fringing reefs are generally being widened by the continuing growth of coral at their seaward margins.

An intermittent fringing reef extends along the coast of Queensland, N of Cairns. The reef platform adjoining the coast is generally
strewn with sand, partly of terrigenous origin, and there are patches
of mangrove swamp bordering the shoreline near high tide mark.
Live coral is confined to the outer edge which is briefly exposed at
low tide.

Several of the high islands off the Queensland coast also have
fringing reefs. They are present on the shores of Lizard Island, on
the islands of the Flinders Group, and on several of the islands in
the Whitsunday Group, notably Hayman Island. Fringing reefs
also occur on the N coast of Australia, notably on the shores of
Arnhem Land, around Darwin, and between Port Hedland and
North West Cape. They are missing from swampy sections of the
shores of the Gulf of Carpentaria and other northern gulfs which
are receiving deposits of land-derived sediment. In these sectors
they give place to tidal flats of the kind described by Russell and
McIntyre (1966).

Offshore reefs
Reefs that lie offshore pose other problems. They include atolls,
barrier reefs (roughly parallel to the coastline and separated from
it by a relatively shallow lagoon), and isolated reef platforms. Each
type is represented in Australian coastal waters. The Great Barrier
Reefs, which extend for about 2000 km from N to S off the Queensland coast, consist of an outer barrier reef (a chain of elongated or
crescentic reefs with intervening gaps, generally about half a mile
wide) in the N, lying about 130 km off the tip of Cape York
Peninsula, about 50 km off Port Stewart, less than 15 km off Cape
Melville, and about 56 km seaward from Cooktown. Off Cairns
there is a broad transverse passage known as Trinity Opening, but
the outer barrier reef reappears farther south, diverging from the
mainland coast until it lies about 100 km off Rockhampton. At the
southern end it breaks up into the scattered reef platforms of the
Capricorn and Bunker groups, with Lady Elliot Island the southern-
most reef, just S of latitude 24°S. It is not clear whether the southern
limit of reef building in these coastal waters is determined by the
temperature of the sea or by the inhibiting effects on coral growth
of the sand that drifts northwards across the sea floor from Fraser
Island (Agassiz, 1893).

The outer barrier plunges steeply (40°–50°) on its seaward side to water more than 1800 m deep off the northern section, shallowing to about 180 m S of the Trinity Opening. Coral is growing to a depth of about 45 m on this seaward slope, which is marked by patterns of grooves and spurs (buttresses) aligned at right angles to the reef edge. These serrations are common on reef margins exposed to strong wave action, but their mode of formation is not clearly understood. The grooves often show evidence of scouring by swash and backwash, and abrasion by waves armed with reef debris, and the spurs are crowned by a rich compact coral and algal growth, but the differentiating mechanism is obscure. At low tide the outer barrier reef is exposed as a chain of elongated platforms, 3–25 km long and up to a kilometre wide, typically with an outer rim on which the surf breaks, a shallow moat, and an algal crest, beyond which the reef slopes gently (5°–10°) into the quieter waters of the lagoon. Yonge Reef (Fig. 91) is a typical outer barrier segment, crescentic in form, with recurves at the N and S ends bordering gaps in the reef. The recurved form is evidently analogous to the recurved spits that border gaps in between barrier islands of sand or shingle, with outlines determined by refraction of waves that enter the gap (Chapter V). The gaps may commemorate the sites of river outlets at low sea level stages. They persist because the rising and falling tide create a strong scouring current which prevents them from being sealed off by coral growth and reef sedimentation.

In one section, SE of Cape Melville, there is a double barrier, with parallel reefs separated by a lagoon about 8 km wide. The inner barrier is less regular in form than the chain of outer barrier reefs, and may be a relic of an earlier barrier reef outflanked by the enlargement of a younger outer reef.

The lagoon between the outer barrier reef and the mainland coast of Queensland is generally between 18 and 45 m deep, with a somewhat featureless floor, because the deposition of land-derived sediment washed in by rivers and reef-derived sediment washed in by the sea has smoothed over any inequalities. Recent work on lagoon-floor sediments in New Caledonia has indicated a high proportion of calcareous organic sediment (typically 80 to 90 per cent) between the coast and barrier reef, with land-derived sediment mainly confined to the zone close to the shore (Guilcher, 1963). The width of a lagoon behind a barrier reef is related to pre-existing sea floor topography and the rate and pattern of upward growth of the reef during Holocene submergence. Coral is growing within the lagoon off the Queensland coast to a depth of about 12 m and columns of reef limestone have grown up from the lagoon floor to form patch reef platforms a little above low tide level. These include the scattered platforms of the Bunker and Capricorn Groups, and the complex pattern of reefs that border the Steamer Channel N of Cairns. The shapes of these reef platforms are related to patterns of waves generated by the prevailing SE trade winds in Queensland coastal waters; several of those that have been built up to low tide level have developed a horseshoe form, with arms trailing northwestwards, away from these prevailing winds (Fig. 87). Cairns Reef has this form, and a more advanced

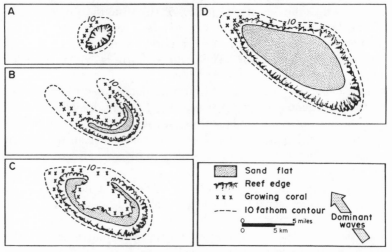

87   *Evolution of horseshoe reefs. Initial reef patches (A), common in coastal waters off Queensland (cf. Fig. 90), tend to grow into forms with marginal deflection by the dominant waves. Stage B is exemplified by Cairns Reef, stage C by Pickersgill Reef, off NE Queensland. Eventually a larger reef platform may develop (D) (after Fairbridge, 1950).*

stage is represented by Pickersgill Reef, the arms of which have curved in almost to enclose a shallow lagoon. In some respects, Pickersgill Reef resembles an incomplete atoll, but the lagoon is only a few feet deep, whereas the atolls commonly found in tropical seas enclose lagoons that are generally at least 35 m deep (Wiens, 1962). Fairbridge (1950) recognised three kinds of true atolls:

oceanic atolls which have localised, generally volcanic, foundations at depths exceeding 550 m, shelf atolls which rise from the continental shelf, with foundations at depths of less than 550 m, and compound atolls where the ring-shaped reef surrounds or encloses relics of earlier atoll formations. Oceanic atolls are common in the W Pacific, and are present in the Coral Sea and the N Tasman Sea, including Middleton and Elizabeth Atolls. Shelf atolls are found off the N coast of Australia, where barrier reefs have not developed (Fairbridge, 1952). Seringapatam, north of Broome, is an excellent example, rising abruptly from a depth of almost 550 m at the outer edge of the broad, sloping continental shelf. The enclosing reef is about 900 m wide, and the lagoon, 9·6 km long and 6·4 km broad, has an average depth of 35 m. Scott Reefs, not far away, consist of a similar enclosed atoll and a second atoll that is incomplete, with the superficial form of a horseshoe reef. Water more than 180 m deep surrounds and separates the two reefs, but the enclosed and partly enclosed lagoons have floors at a depth of 35–45 m. Compound atolls are

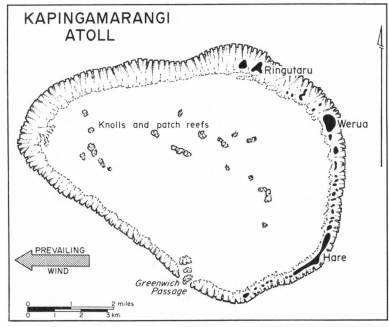

88   *Kapingamarangi Atoll, Caroline Islands (after Wiens, 1962)*

represented by the Houtmans Abrolhos Group, which consists of a
complex of reef platforms, with remnants of emerged reefs of
Pleistocene age (Fairbridge, 1948). Typical features of an oceanic
atoll are shown in Fig. 88. The reef flat on the windward side is
often wider, with a *Lithothamnion* rampart in the breaker zone, and
sometimes one or more low islands of coral debris. Outlets to the
ocean are usually on the leeward side. The lagoon floor is generally
a smooth depositional surface, from which pinnacles and ridges
of live coral may protrude (Fig. 89). There are, however, many
variations as indicated by the many detailed studies of atolls
recorded in the *Atoll Research Bulletin*.

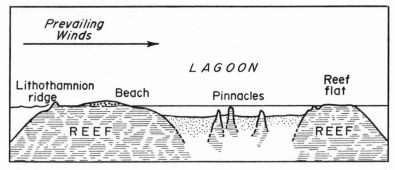

89   *Diagrammatic section across a typical atoll*

Origin of reef forms

The central problem of reef development is the upward growth of
reefs from below depths at which the reef-building corals now live.
The foundations of barrier reefs and atolls, if not of fringing reefs,
are well below the present limit of coral growth, and yet the
formation of coral structures has been essential for their construction. In 1837 Charles Darwin observed reef landforms during his
voyage across the Pacific in the *Beagle*, and proposed the subsidence
theory, which regarded barrier reefs and atolls as the outcome of
upward growth from ancient fringing reefs bordering islands and
continental margins that have since subsided beneath the oceans
(Darwin, 1842). In these terms, the outer barrier reef off the E
Queensland coast commemorates the alignment of a former coast
that has subsided tectonically. Atolls, some of which still have a
central island ('almost-atolls'), are regarded as the built-up fringing
reefs encircling islands lowered by crustal subsidence: an island

with a fringing reef subsides to become an island surrounded by a barrier reef and enclosing a lagoon, and finally an atoll, with the central island lost from view (Fig. 86). The reefs we see are those that have been able to maintain upward growth as subsidence took place. Those that have been carried down too rapidly are now submerged beneath the depth at which reef-building organisms can live. Submerged barrier reefs and atolls have also been detected by soundings in the ocean particularly off the SE coast of New Guinea and in the Coral Sea. Many of the sea-mounts which rise from the deep floor of the Pacific Ocean have been planed off to form flat-topped guyots with surfaces at depths of 550–900 m, well below the possible limits for Quaternary reef initiation. Some bear fossil corals of Cretaceous age, and have evidently subsided below the limit of reef growth during Tertiary and Quaternary times. Platforms that remained relatively shallow provided the foundations for the existing pattern of surface and near-surface reefs. Some anomalies remain. It is difficult to understand why the coral is dead on Alexa Bank, a drowned atoll NW of Fiji at a depth of 18–27 m, in an environment apparently suitable for reef growth. It is possible that the coral was killed during a low sea level phase, and that there has been a failure in recolonisation by living reef-builders; that the site has been damaged by toxic materials, possibly volcanic material; or that there has been an upwelling of foul muds or cold water from adjacent ocean depths (Fairbridge and Stewart, 1960). None of these hypotheses is really satisfactory. In general, islands and reefs that were exposed during Pleistocene low sea level phases retained live marginal corals as nuclei from which a reef-building community developed during episodes of sea level rise.

The subsidence theory is an attractive one, and it has received support from many later workers. Davis (1928) pointed out that coasts bordered by barrier reefs, such as the E Queensland coast, show abundant evidence of submergence by the sea, in the form of inlets, embayments, and drowned valley mouths. He regarded the general absence of cliffing on such coasts as evidence that the barrier reefs grew up as the continental shelf subsided, so that the coast was consistently protected from the action of strong ocean waves. This would explain the contrast between the E Queensland coast, with its numerous promontories and 'high islands' offshore, and the New South Wales coast, farther S and beyond the protection of the barrier reefs. On the E Queensland coast, cliffing is clearly related to waves generated by the prevailing SE trade winds in coastal waters behind the Great Barrier Reefs. On the New South Wales coast, where submergence is equally in evidence, and there are fewer offshore islands, the headlands are more strongly cliffed. It is evident that the E Queensland coast has never been exposed to the full force of ocean waves that have shaped the coastal landforms of New South Wales, and the building of a barrier reef during a phase of subsidence provides a possible explanation for this. Where coastlines bordered by barrier reefs have fringing reefs as well (as in NE Queensland) it is necessary to regard the latter as secondary forms, initiated along the coast after the subsidence which led to the upgrowth of the barrier reef came to an end (Davis, 1928).

A second theory, more easily discredited, is the antecedent platform theory proposed originally by Murray (1880) and supported by Agassiz (1903). According to this theory, reef-building began when corals and associated algae colonised submarine platforms formed either by marine planation in coastal waters or by accumulation of submarine banks of sediment. Lagoons are regarded as hollows excavated in the coral mass as the result of solution by waves washing over the edge of the reef. The theory requires no changes of land or sea level, and it is assumed that the present relationship between land and sea levels has been in existence for a long time. However, there is no satisfactory evidence in support of any of these suggestions, and modern attempts to explain reef forms are made in the knowledge that land and sea levels have certainly changed in the past.

Many of the features that Darwin ascribed to tectonic subsidence can equally well be explained by a rising sea level, and, with this in mind, Daly (1910, 1934) proposed a glacial control theory to account for reef development. Major glacio-eustatic changes of sea level have taken place during Pleistocene and Recent times (Chapter III), and when sea level fell, possibly as much as 140 m during the Last Glacial phase, any reefs previously formed must have been exposed to the atmosphere and eroded. As ocean temperatures also fell, the growth of coral could have continued only in the warmer parts of the oceans, probably in latitudes to 10°–15° N and S. The present patterns of reef landforms are thus regarded as the outcome of reef building during and since the

Holocene marine transgression, and the barrier reefs and atolls we see are those that grew rapidly enough to keep pace with the rising sea, or at least to remain within the zone where ecological conditions permit coral and algal growth. The narrowness (400–800 m) of many reefs could be a function of relatively rapid upward growth, of the order of 120 m in the last 20,000 years (cf. Fig. 11). In support of his hypothesis, Daly claimed that depths of lagoons enclosed by reefs showed a considerable degree of accordance, few being more than 75 m deep. They originated, he suggested, as reef platforms planed off by the sea when it stood at a low level, When the marine transgression began, live corals at the margins of the planed-off surface developed an encircling reef, which grew upwards as sea level rose. Fringing reefs, in these terms, have developed only since the Holocene marine transgression established the general outlines of the present coast, although some of them probably embody the relics of Pleistocene fringing reefs which developed at an earlier stage close to the present shore level, and escaped destruction during low sea level phases.

Further evidence of the origin of atolls and barrier reefs has been obtained from deep borings in reef structures. In 1926 a boring at Michaelmas Cay (off Cairns) proved reef material to a depth of 123 m, resting upon unconsolidated sand, and a similar boring at Heron Island (Fig. 90) in 1937 penetrated 154 m of reef material before entering sand. In 1959 a much deeper boring on nearby Wreck Island enabled samples to be dated from their content of fossil foraminifera (Derrington, 1960). The top 121 m consisted of Recent reef debris, beneath which Pleistocene sand extended to 162 m, Tertiary deposits, including ancient fragments of coral, to 422 m, then Tertiary sandstone and siltstone to 547 m, where older volcanic basement rocks were encountered. On Bikini Atoll in 1948, borings proved reef material and atoll sediment to about 750 m, and a seismic refraction survey indicated a further 1500–3000 m of material with similar physical properties, possibly more reef debris or sediment of volcanic origin (Dobrin *et al.*, 1949). The reefs are built of shallow water coral and associated organisms and, as their foundations lie deeper than any possible glacio-eustatic lowering of sea level, tectonic subsidence of the ocean floor is required to explain them. A more detailed analysis of the reef stratigraphy deduced from borings would probably yield information on stages in the growth of reefs, and results from further

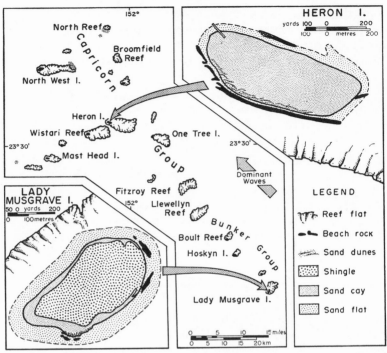

90 *Island reefs of the Bunker and Capricorn groups, off S Queensland, showing Heron Island (a sand cay) and Lady Musgrave Island (a shingle cay) (after Steers, 1937)*

borings are awaited. It is already clear that broad-scale epeirogenic subsidence has taken place on the floor of the Pacific during and since Tertiary times, but there have also been world-wide eustatic oscillations of sea level in Quaternary times, and it is likely that the reefs we now see were built up to their present-day level during and since the Holocene marine transgression, probably plastered on to the dissected stumps of earlier reef formations that developed during high interglacial stands of sea level during Pleistocene times.

A satisfactory explanation of reef landforms evidently requires a combination of Darwin's subsidence theory (to account for the depth of reef foundations) and Daly's glacial control theory (since sea levels have certainly risen and fallen during the period of reef building). The allocation of the respective roles of sea level rise, sea floor subsidence, the provision of foundations suitable for reef building, and the extent of ecological conditions favourable to reef growth, may be possible when more detailed studies have been made of the distribution and stratigraphy of these reef structures.

The distribution of reefs raises a number of questions. On the world scale the pattern reflects the distribution of coral larvae by ocean currents, the paucity of reefs in the E Atlantic and Pacific areas being related to the westward flow of equatorial currents, and the richness of the reefs off Queensland being related to abundant larval supply in currents arriving from the W Pacific and the Coral Sea. It is more difficult to account for the poor development of coral reefs off the W coast of Australia, where the reefs parallel to the mainland coast are submerged ridges of aeolian calcarenite with only a veneer and fringe of living coral and calcareous algae, similar to the sandstone reefs off NE Brazil. Fairbridge (1967) suggests that here, too, the supply of coral larvae has been poor, both the northward summer current and the southward winter current arriving from reefless or relatively reefless ocean areas. The prevailing offshore winds may have helped keep away the coral larvae, and a further limiting factor may be the relatively low nutrient status of the sea off the low-lying, semi-arid coast of Western Australia compared with the steep, high-runoff coast of E Queensland. The patchy development of reefs off N Australia may be due to excessive subsidence in the Timor Sea. A submerged barrier reef exists off the Sahul Shelf, indicating that sea floor subsidence has been too rapid for reef growth to be maintained.

Emerged reefs

The fact that reef platforms are built up to a strict level in relation to low tide makes them good indicators of emergence where they are found at a higher level, but it is difficult to decide whether this emergence is the result of tectonic uplift or of eustatic lowering of sea level. The latter explanation is preferred by Fairbridge for the emerged reef that forms part of the Pelsart Group in Houtmans Abrolhos, off the coast of Western Australia, which was evidently built to the level of a Late Pleistocene interglacial (or interstadial) stillstand of the sea about 7·5 m above present sea level. It is bordered by modern reef platforms, which are still developing just above low tide level (Fairbridge, 1948). Similar emerged Pleistocene reefs have been found farther S, on Rottnest Island, beyond the present range of reef-building (Teichert, 1950).

Emerged reefs extending up to 3 m above present low tide level have been reported at various places around the Australian coast, including the shores of Melville Island, the reefs in Torres Strait, and several islands off the Queensland coast, notably Raine Island. They have generally been regarded as reef platforms built up to a slightly higher sea level, then laid bare as a consequence of a succeeding drop in sea level. Steers (1937) reported emerged reef

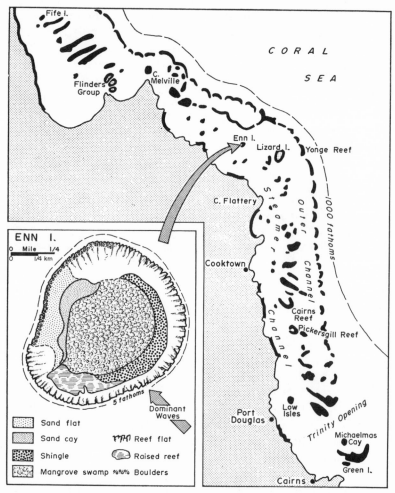

91   *Reefs off NE Queensland. Inset: Enn Island, a low wooded island (after Steers, 1937).*

platforms between 1·5 and 3 m above low tide level from islands off the NE coast of Queensland: for example, on Enn Island (inset, Fig. 91), where a raised reef 2–2·75 m above the present reef platform is being undercut marginally by wave action. Benches have been observed at a similar level, cut in older non-coralline rocks on the shores of islands off this coast, and also on the flanks of headlands, notably Cape Flattery (Steers, 1937). In his summary analysis of the reef landforms of Australian coastal waters, Fairbridge (1950) found evidence of emerged reef platforms corresponding with the sequence of Recent higher sea levels he had previously inferred from terraces cut in aeolian clacarenite at 3, 1·5 and 0·5 m above present low tide level on the coast of Western Australia. The validity of these Recent higher sea levels was discussed in Chapter III; if the views of Shepard and others that the Holocene marine transgression attained present sea level without ever exceeding it gains acceptance, it will be necessary to find some other explanation for these emerged reef platforms. Two possibilities remain. The reef platforms may have been built up to higher sea levels during Pleistocene times, but if they are really of Recent age they must have been raised out of the sea by tectonic movements during the past few thousand years.

The intricacies of the problem are well illustrated by the question of the emerged reef that borders Peel Island, in Moreton Bay (27°30′ S), S of the present limit of coral growth in Queensland coastal waters. It is tempting to regard this as a reef platform of Recent formation, built up to the surface of a higher (and probably warmer) sea when the Holocene marine transgression attained a maximum level. A radiocarbon date for a sample of dead coral placed it at 3710±250 years B.P. (Gill, 1961), but critics have questioned the validity of datings of emerged coral. The Peel Island coral may owe its emergence to localised tectonic uplift. Alternatively, it may be a reef platform that was built up in Pleistocene times to the level of a higher interglacial or interstadial sea.

Examples of emerged reefs that have certainly been elevated by tectonic movements are found on the mountain slopes of the N coast of New Guinea where reefs extend more than 1500 m above sea level, far above the levels attained by any previous eustatic movements of the sea. They rest unconformably on an uplifted sea floor, and are being rapidly consumed by subaerial processes of denudation, notably solution by rainwater. On the N coast of the

Huon peninsula a stairway of emerged reefs forms terraces which diverge laterally across an axis of upwarping, attaining a maximum elevation of about 750 m in the vicinity of the Tewai gorge. Russell (1966) has described a similar sequence of uplifted fringing reefs on the tilted coral cap of the island of Barbados. Christmas Island, 320 km S of the coast of Java, is an emerged atoll 390 m high, uplifted tectonically to such an extent that underlying Eocene marine limestones are exposed. The Loyalty Islands, NE of New Caledonia, include uplifted atolls such as Mare, where the former lagoon floor is an interior plain surrounded by an even-crested rim of reef limestone 30 m high, and Uvea, which has been intermittently tilted during upheaval so that the terraced eastern rim has emerged while the western rim, submerged, has a newer loop of developing reefs. Emerged reef landforms such as these are being attacked marginally by marine erosion processes, with the result that low tide shore platforms are being developed similar to those that border Australian calcarenite coasts (Chapter IV). The Isle of Pines, S of New Caledonia, is fringed by emerged reefs with typical notch-and-visor coastal profiles, and a shore platform that is being extended seaward by the addition of a modern fringing reef.

Evidence of lower stillstands of sea level is less easily obtained from within reef landforms, but platforms developed during such episodes may be indicated in borings by unconformities, weathered layers, or horizons of reef sediment. That reef platforms are quickly responsive to sea level changes is implicit in Fairbridge's (1947) observation of the renewed upward growth of coral on reefs in Houtmans Abrolhos, which he regarded as a consequence of the contemporary worldwide rise of sea level indicated by tide gauge records and other evidence (Chapter III).

Islands on reef platforms

In addition to islands formed by the emergence of reefs, various kinds of 'low island' have been built up on reef platforms by accumulation of sand, shingle, and boulders formed from reef debris that has been eroded by wave action and cast up on the platform. Some of these low islands are built of sediment derived from the destruction of emerged reefs, but many are not, and it is clear that although emergence would facilitate their development, it is by no means essential, for they can be built without any change in the relative level of the sea. A variety of carbonate sediments make up the depositional features on these islands, including large lumps of broken reef material, sticks of staghorn coral, discoidal *Halimeda* shingle, the grits and sands into which coralline material disintegrates, and fine-grained carbonate muds (Folk and Robles, 1964).

The simplest kinds of low island are exemplified by the patch reef platforms of the Bunker and Capricorn groups off the Queensland coast (Fig. 90). The prevailing winds here are the SE trades, and the islands have generally been built near the NW corner of the reef platform, because debris eroded from the reef is washed across the platform by waves from the SE at high tide. Refraction of waves around the reef platform produces a convergence on the lee side, building up waves in such a way that they prevent the reef debris from being swept over the lee edge of the platform. At first the island is nothing more than a sandbank or a heap of coral shingle awash at high tide, but gradually the sediment accumulates, and the island is built above high tide level, colonised by grasses and shrubs, and then by trees, notably palms (*Pisonia* and *Pandanus* species), and *Casuarina*. An island of this type is termed a cay. It is often elongated at right angles to the prevailing winds, but its configuration is subject to change as erosion and deposition alternate on its shores.

Between high and low tide levels, sand and shingle are often cemented by secondary deposition of calcium carbonate in the zone of repeated wetting and drying. The sand forms the compact sandstone layer known as beach rock (Plate 32), described in Chapter V, whereas the shingle becomes a lithified conglomerate. The greater resistance of these formations stabilises a cay to a certain extent, although waves generated during tropical typhoons may still sweep it away, or modify its outline, leaving only the patterns of eroded beach rock or shingle conglomerate to commemorate its former configuration.

The Bunker and Capricorn groups are not protected from ocean waves by any barrier reef. On the Bunker Group, exposure to ocean waves is evidently responsible for the predominance of coarse shingle over sand in such islands as Lady Musgrave Island (Fig. 90), while on the less exposed platforms of the Capricorn Group, sand cays such as Heron Island (Fig. 90 and Plate 33) are more common. Both these islands show evidence of slight changes of configuration, their former outlines being marked by

33 *An aerial view of Heron Island, a sand cay on a reef platform off the coast of S Queensland (L. & D. Keen Pty Ltd)*

34 *Low Isles, a low wooded island off Port Douglas, Queensland, showing the shingle rampart on the SE shore in the foreground, the mangrove swamp, the lagoon, and the sand cay on which the lighthouse stands*

lines of beach rock. Recession of the S shore of Heron Island has left a ridge of beach rock offshore, and wind action has built some of the eroded coral sand into low dunes.

More complex low islands are found off NE Queensland, on reef platforms in the lagoon N of Cairns (Fig. 91). In addition to sand cays similar to Heron Island (Fife Island, N of Port Stewart, and Green Island, off Cairns, are examples of these), there are islands consisting of a sand cay on the leeward (NW) side, a shingle embankment on the windward side, and an intervening depression, or shallow lagoon, in which mangrove swamps (chiefly the red mangrove *Rhizophora*, but some of the smaller white mangrove, *Avicennia*, marginally) often develop. These are known as 'low wooded islands' (Figs. 92, 93), and Enn Island (Fig. 91) and Low

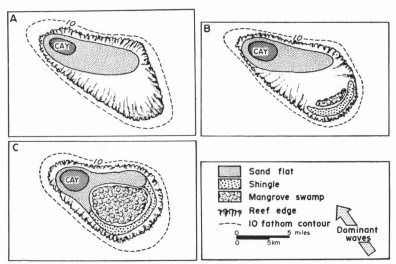

92   *Evolution of a low wooded island on a reef platform. A, platform with sand cay (e.g. Fife Island); B, addition of shingle ridges on seaward side; C, development of a mangrove swamp (e.g. Enn Island; Low Isles).*

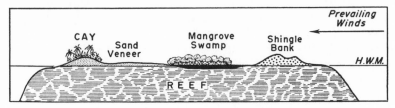

93   *Section across a low wooded island on a reef platform*

Isles (Plate 34) are good examples. Their development is evidently related to the greater shelter afforded by the outer barrier reef here, compared with the Capricorn and Bunker groups. Sand cays develop first, on the leeward side, and then shingle ridges are thrown up on the windward side, built from debris eroded from the edge of the reef platform. Mangroves then colonise the inter-vening depression and there is evidence that they 'dig themselves in' as organic acids exuded from the roots corrode the underlying reef limestone, replacing it with soft carbonaceous mud. In some cases there are several shingle ridges; the older ridges are dark in colour and largely cemented as conglomerate, but the younger are white or cream, and relatively loose.

It has been suggested that some of these islands include plat-forms cut across shingle conglomerate when the sea stood at a relatively higher level (Steers, 1929), but there are difficulties with this hypothesis that have never been clearly resolved; it is uncertain, for instance, how and when the shingle was placed on the reef. If it was built up and cemented on the reef platform with the sea at its present level, it is necessary to explain why the reef was not built up subsequently when the sea stood at the higher level claimed for planation of the conglomerate. Enn Island embodies an emerged reef, which could have been built up to a higher Recent sea level, but could also be an inheritance from Pleistocene times, or uplifted tectonically. It is conceivable that much of the coral shingle on these low wooded islands was formed as wave action destroyed reefs that had previously been built up, or lifted tectonically, above present sea level, but Stoddart (1965a) has shown that it is not necessary to invoke Recent emergence, either tectonic (Spender, 1930) or eustatic (Steers, 1937), to account for the distribution of cays and low wooded islands off the Queensland coast. They are related to wave energy conditions. On the outer reefs, in a high wave energy environment, these islands are not formed. The inner reefs, partly protected from ocean waves, have a limited fetch and a medium wave energy environment conducive to sand cay forma-tion by refracted waves generated by local winds. The reef plat-forms in the Steamer Channel are in deep water with a relatively broad fetch over which the SE trade winds can generate waves strong enough to build shingle ramparts on the windward side and cays on the leeward side, as in the low wooded islands, but not so strong as to sweep the reefs clean as on the outer barrier. Similar patterns of cays and low wooded islands occur in similar conditions of wave energy variation in the reefs off the British Honduras coast (Stoddart, 1965a).

It is clear that these low islands are by no means stable features, for there can be little doubt that during exceptional storms they are much reduced and may even be swept completely off the reef platform. Their reconstruction in similar, if not identical, patterns is a relatively slow process; it can be studied by making repeated surveys of particular islands. On Low Isles, repeated surveys have shown changes in configuration resulting from accretion during relatively calm conditions and erosion during severe storms (Spender, 1930; Steers, 1937; Fairbridge and Teichert, 1948). Stoddart (1965b) made a detailed study of the effects of Hurricane Hattie (October 1961) on the atolls and barrier reef islands off the British Honduras coast, comparing their subsequent configuration with surveys of their physiography and vegetation made in 1959–61. The hurricane swept westwards across the reefs towards the coast S of Belize, and was accompanied by gusts exceeding 320 km/hr. A 72 km-wide storm surge attained up to 4·5 m above normal high water. Reef islands which had a dense littoral vegetation were modified less than those on which the natural vegetation had been cleared and replaced by coconut plantations: the former were eroded on the windward side with shingle and coral debris piled up against the vegetation, but the latter were more severely damaged, and seven cays were swept away completely. Stoddart estimated that the natural restoration of damaged areas would take 25–30 years.

In many respects these low islands are similar to the spits and barriers built by coastal deposition as described in Chapter V; their distinctive features are largely a response to the special environment of the reef platforms on which they have been built.

X

CLASSIFICATION OF COASTAL LANDFORMS

Various attempts have been made to classify coastal landforms (including shores and shoreline features), but none of them have been entirely successful. This is partly because of a widespread belief in the superiority of genetic classifications, in terms of the origin of the landforms, over purely descriptive classifications, in terms of landforms as observed: cliffed coasts, fiord coasts, mangrove coasts, and so on. The difficulty is that a genetic classification can only be applied satisfactorily when the mode of origin of coastal landforms is known; the attempt to use particular types or associations of landforms as indicators of particular modes of origin has frequently led to errors that have been revealed by subsequent coastal research.

One of the early attempts at coastal classification was that put forward by Suess (1906). He distinguished 'Atlantic' coasts, which run transverse to the general trend of geological structures, from 'Pacific' coasts, which run parallel to structural trends. The former are characteristic of the Atlantic shores of Europe; the latter of the Pacific coast of North America. In Australia, the Atlantic type is exemplified by the coast between Broome and Darwin, particularly around Yampi Sound and the Buccaneer Archipelago (Fig. 94), and the Pacific type, appropriately, by the Pacific coast of New South Wales, which is parallel to the N-S strike of the rocks of the Eastern Highlands. This classification is too broad to be of much use in geomorphological studies.

Another method of classification distinguishes coasts formed by submergence from coasts formed by emergence. This distinction was made by Gulliver (1899), and developed into a genetic classification of *shorelines* by Johnson (1919), who proposed four categories:

1. *Shorelines of submergence*

2. *Shorelines of emergence*

3. *Neutral shorelines*, with forms due neither to submergence nor emergence, but to deposition or tectonic movement: delta

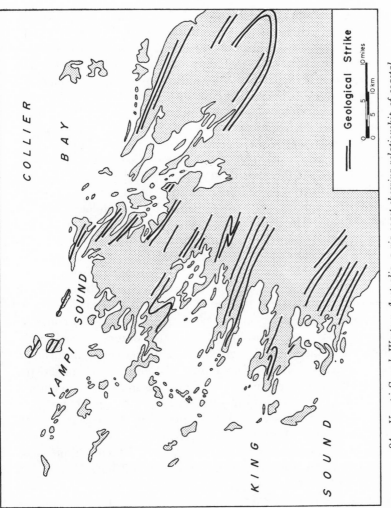

94 *Yampi Sound, Western Australia, a ria coast showing relationship of coastal configuration to geological strike (after Edwards, 1958)*

shorelines, alluvial plain shorelines, outwash plain shorelines, volcano shorelines, and fault shorelines.

4. *Compound shorelines*, which have an origin combining two or more of the preceding categories.

There are difficulties in applying this classification, particularly if it is restricted to shorelines as previously defined ('the water's edge, migrating to and fro with the tide'). Johnson apparently took a wider view than this, including offshore barriers as shoreline features, and his scheme may be more conveniently considered as a classification of coasts. Difficulties remain, however, for most coasts show evidence of both emergence, following high sea levels in interglacial phases of the Pleistocene period, and submergence, due to the Holocene marine transgression. Many show 'neutral' features as well, so that most fall into the 'compound' category. Johnson applied the classification in terms of the most strongly marked characteristics; the majority of coasts must fall into the category of submerged coasts, shaped by the Holocene marine transgression. The possibility of a slight emergence following this transgression (Chapter III) should be borne in mind, but strongly marked characteristics of emerged coasts can really be expected only in regions that are rising because of isostatic recovery following deglaciation, as in Scandinavia, or in regions subject to uplift associated with tectonic deformation, as in New Zealand.

Johnson suggested certain characteristics of 'shorelines of submergence', such as an initially indented outline formed by the drowning of valley mouths, which are readily recognised from maps, nautical charts, and air photographs. The proposed characteristics of 'shorelines of emergence', initially simple and smooth in outline, with barriers likely to form offshore where the depth has been reduced by emergence, evidently apply only to coasts of gentle offshore gradient. Emergence of steep coasts with irregularities in the sea floor offshore could produce a more indented outline, as on the Californian coast (Putnam, 1937), but offshore barriers are certainly not restricted to 'shorelines of emergence': they are most common on recently-submerged coasts. While Johnson did not insist that these characteristics were diagnostic of 'shorelines of emergence' the exceptions show that application of this genetic classification is hazardous until coasts have been investigated in sufficient detail to determine their geomorphological history.

In 1899 Gulliver proposed a distinction between initial and subsequent coast forms, the initial forms to be those that existed when the present relative levels of land and sea were established and marine processes began work, and the sequential forms to be those that have since developed as the result of marine action. Shepard (1937, 1948) devised a more precise classification on these lines, which made a basic distinction between coasts shaped primarily by non-marine agencies and coasts that have been modified to their present form by the activity of marine processes. It was essentially a genetic classification, with descriptive detail inserted to clarify the subdivisions, and it recognised that, because of the world-wide Holocene marine transgression, the sea has not long been at its present level relative to the land, and many coasts have been little modified by marine processes. His aim was to devise a classification that would 'prove useful in diagnosing the history and origin of coasts and shorelines from a study of charts and airplane photographs', but it is dangerous to assume that the history and origin of a coast can be deduced from such evidence without a field investigation. A straight coast may be produced by deposition, faulting, emergence of a featureless sea floor, or submergence of a coastal plain; an indented coast may be formed by submergence of an undulating or dissected land margin, emergence of an irregular sea floor, differential marine erosion of hard and soft outcrops at the coast, or transverse tectonic deformation (folding and faulting) of the land margin. It is doubtful whether configuration can be accepted as a reliable indication of coastal evolution.

Shepard's (1948) proposed classification of coasts (including shorelines) is as follows:

I. Primary or youthful coasts with configuration due primarily to non-marine agencies:

   A. Shaped by subaerial denudation and submerged by the Recent marine transgression or by down-warping of the land margin:
      1. Drowned mouths of river valleys (ria coasts)
      2. Drowned mouths of glaciated valleys (fiord coasts)
   B. Shaped by subaerial deposition:
      1. River deposition:
         a. Deltaic coasts
         b. Alluvial plain coasts

2. Glacial deposition:
   a. Partially-submerged morainic coasts
   b. Partially-submerged drumlin coasts
3. Wind deposition: dune coasts
4. Vegetation extending to coast: mangrove coasts

C. Shaped by volcanic activity:
   1. Volcanic deposition (e.g. lava flow) coasts
   2. Concave embayments formed by volcanic explosion

D. Shaped by earth movements:
   1. Coasts produced by faulting
   2. Coasts produced by folding

II. Secondary or mature coasts with configuration primarily the result of marine agencies:

A. Shaped by marine erosion:
   1. Cliffed coasts made more regular by marine erosion
   2. Cliffed coasts made less regular by marine erosion

B. Shaped by marine deposition:
   1. Coasts made more regular by marine deposition
   2. Coasts made less regular by marine deposition
   3. Coasts with offshore barriers
   4. Coasts with 'coral reefs'

In 1963 Shepard modified and elaborated this classification, but retained its basic structure. The chief modifications under Primary coasts were the addition of a category, drowned karst topography (I-A-3), the elaboration of wind-deposition coasts (I-B-3) to include fossil dune and calcarenite coasts, the deletion of mangrove coasts from (I-B-4) and the substitution there of landslide coasts, and the addition of a new category (I-D-3) of coasts produced by sedimentary extrusions, notably salt domes and mudlumps. Under Secondary coasts, category B, coasts shaped by marine deposition, was revised thus:

II. B. Shaped by marine deposition:
     1. Barrier coasts
     2. Cuspate forelands

3. Beach plains
4. Mud-flats or salt marshes

and an additional category C added:

II. C. Built by organisms:
1. Coral reef coasts
2. Serpulid reef coasts
3. Oyster reef coasts
4. Mangrove coasts
5. Marsh grass coasts

The 1948 scheme and the 1963 revisions have both been given here, because they illustrate the difficulties of achieving a comprehensive classification. The chief difficulty is to decide at what stage a coast has been sufficiently modified by marine processes to place it in Group II rather than Group I. Many coasts fall into more than one category, and may even overlap the basic division: barrier coasts (II-B-1) owe many of their features to deposition by wind (I-B-3), and now that calcarenite coasts have been included in the wind deposition category (I-B-3) there is a difficulty when they have been cut back and cliffed on the seaward side—made less regular by marine erosion (II-A-2). The 1963 version incorporates additional descriptive categories under genetic headings, and the implication of a particular mode of origin raises difficulties. The distinction between salt marshes (II-B-4) and marsh grass coasts (II-C-5) is subtle: the first category implies deposition of sediment followed by vegetation colonisation, the second deposition of sediment caused by vegetation colonisation, but, if this distinction is made, it raises a question as to whether mangrove coasts (II-C-4) should additionally or alternatively be represented under category II-B.

Examples of most of Shepard's categories have been mentioned in this book. Ria coasts and fiord coasts are noted in Chapter VII, and deltaic coasts in Chapter VIII. Alluvial plain coasts, formed by deltaic deposition from a group of neighbouring streams, are exemplified from the E coast of South Island, New Zealand, partially submerged morainic coasts from Long Island, and partially submerged drumlin coasts from Boston Harbour. Coasts literally formed by wind deposition must be rare, confined to sheltered lagoon or embayment shores where a dune has spilled over a barrier or an adjacent promontory: there are examples on the S shore of Corner Inlet, in SE Australia. The Axmouth-Dowlands landslip in E Devon (Chapter IV) is a landslide coast, and lava flow coasts are well developed W of Mauna Loa, an active volcano in Hawaii. Embayments formed by volcanic collapse or explosion are rare: Port Lyttleton, the harbour for Christchurch, was once thought to have originated in this way, but it is now known to be a drowned inlet, formed by marine submergence of a river valley incised into an ancient volcano (Cotton, 1952). Folded and faulted coasts occur in the Wellington district, New Zealand. Salt domes form oval-shaped islands in the Persian Gulf and mudlumps are found off the shores of the Mississippi delta.

Examples of coasts with configuration due primarily to marine action include the chalk coasts of the English Channel, which are irregular in detail but have developed a broadly-smoothed outline in plan, and the cliffs of Port Campbell, Victoria, which have been made more irregular by marine erosion. Barrier coasts are described in Chapter V, along with cuspate forelands and prograded beach plains, which are similar to barriers but adjoin the coast without any intervening lagoon or swamp. Mud flats, salt marshes, and mangrove coasts are exemplified in Chapter VII, and coral reef coasts in Chapter IX. Shepard does not quote examples of serpulid reef coasts, oyster reef coasts, or marsh grass coasts. He comments that his classification has been used successfully in class work, and its applications in devising laboratory exercises are obvious, but, in terms of coastal research, a classification is useful if it provides a series of descriptive categories requiring investigation and explanation. There are dangers when descriptive categories based on interpretation of maps or aerial photographs are fitted into a pre-arranged genetic framework; in successive revisions Shepard has increased the descriptive categories and made his scheme as comprehensive as is possible while remaining within the limitations of a genetic approach.

An omitted category which has recently been receiving closer attention is the category of ice coasts, well developed in Antarctica, where ice shelves terminate in cliffs up to 35 m high. In summer, icebergs calve from these and in winter the adjacent sea freezes, a fringe of sea ice persisting until winds and the spring thaw break it into floes. Melting in late spring is hastened by diatomaceous growths which darken the ice and increase its solar heat absorption. Protruding snouts of glaciers which reach the sea form related features, somewhat richer in morainic debris, which is released and deposited on the adjacent sea floor. Law (1967) gives a brief description of these coastal features, but further work is necessary to determine their morphogenic relationships.

Of the several classifications that have been proposed by Cotton, the 1952 scheme is the most straightforward (Cotton, 1952). Aware of the importance of tectonic movements in shaping the coasts of his homeland, New Zealand, he made a fundamental distinction between coasts of 'stable' and 'mobile' regions. 'Stable' regions have not been affected by tectonic movements during Quaternary times, while 'mobile' regions have been subject to Quaternary earth movements, which may still be continuing. The classification of coasts bordering these regions is as follows:

A. Coasts of stable regions. These have all been affected by the Recent marine submergence:
   1. coasts dominated by features produced by Recent submergence
   2. coasts dominated by features inherited from earlier episodes of emergence
   3. miscellaneous coasts (fiord, volcanic, etc.)

B. Coasts of mobile regions. These have been affected by uplift or depression of the land as well as by Recent submergence:
   1. coasts on which the effects of marine submergence have not been counteracted by uplift of the land
   2. coasts on which recent uplift of the land has led to emergence
   3. fold and fault coasts
   4. miscellaneous coasts (fiord, volcanic, etc.)

This system is most useful as a clarification of Johnson's scheme, the analysis of submergence and emergence enabling certain types of 'compound' coast to be separated.

Another approach to coastal classification is that proposed by Valentin (1952; see also Cotton, 1954), who made a fundamental distinction between advancing and retreating coasts, noting that advance may be due to coastal emergence and/or progradation by deposition, while retreat is due to coastal submergence and/or retrogradation by erosion. He first devised a system of coastal classification that could be used on a world map (scale 1:50,000,000) of coastal configuration types (Valentin, 1952, plate 1):

A. Coasts that have advanced:
  1. due to emergence:
     a. *emerged sea floor coasts*
  2. due to organic deposition:
     b. phytogenic (formed by vegetation): *mangrove coasts*
     c. zoogenic (formed by fauna): *coral coasts*
  3. due to inorganic deposition:
     d. marine deposition where tides are weak: *lagoon-barrier and dune-ridge coasts*
     e. marine deposition where tides are strong: *tideflat and barrier-island coasts*
     f. fluvial deposition: *delta coasts*

B. Coasts that have retreated:
  1. due to submergence of glaciated landforms:
     g. confined glacial erosion: *fiord-skerry coasts*
     h. unconfined glacial erosion: *fiard-skerry coasts*
     i. glacial deposition: *morainic coasts*
  2. due to submergence of fluvially-eroded landforms:
     k. on young fold structures: *embayed upland coasts*
     l. on old fold structures: *ria coasts*
     m. on horizontal structures: *embayed plateau coasts*
  3. due to marine erosion:
     n. *cliffed coasts*

The classification is partly genetic and partly descriptive, and examples of each type of coast shown on Valentin's world map are listed below:

  a. *emerged sea-floor coasts* border the S shores of Hudson Bay, Canada, where the land is emerging as the result of isostatic recovery following deglaciation of the Canadian Shield.
  b. *mangrove coasts* are found in sheltered embayments on tropical coasts. On the world map they were shown to extend from Broome, on the N coast of Western Australia, eastwards to Cape York, then S almost to Cooktown but, as Valentin

(1961) has since demonstrated, they occur only on restricted sections of this coast, chiefly around estuaries.

c. *coral coasts* are exemplified by the Queensland coast, bordered by the Great Barrier Reefs.

d. *lagoon-barrier and dune ridge coasts* are well developed on the SE coast of Australia and the Gulf and Atlantic coasts of the United States.

e. *tideflat and barrier-island coasts* are found on the Dutch, German, and Danish North Sea coasts.

f. *delta coasts* are coasts with major deltas, such as the Mississippi and the Nile.

g. *fiord-skerry coasts* are best developed in Norway and British Columbia; the W coast of New Zealand's South Island has fiords, but skerries (fringing islands) are relatively rare. Port Davey, in Tasmania, shown as a fiord by Valentin, is actually a drowned river valley.

h. *fiard-skerry coasts* are found on the Baltic shores of Sweden and Finland and on the Arctic coasts of Canada.

i. *morainic coasts* are exemplified by the S Baltic coast.

j. *embayed upland coasts* are found on the W coast of the United States, in Chile, and in the Mediterranean, notably in Greece and Turkey.

k. *ria coasts* are represented by S Ireland, SW England, Brittany, and NW Spain, and by the coasts of Tasmania, New South Wales, and S Queensland.

l. *embayed plateau coasts* border the Red Sea and W India, Argentina, and parts of N Russia; the Hampshire coast in S England is placed in this category.

m. *cliffed coasts* are extensive on the Nullarbor Plain margin in S Australia, in parts of S and W Africa, and locally on the shores of the Mediterranean, the Black Sea, the English Channel, and the North Sea.

Valentin's classification has the merit of taking account of changes in the relative levels of land and sea, and being based on observed evidence of the gain or loss of land. He also considered the evidence of changes in progress, which can be expressed as an interaction of vertical movements (emergence and submergence) and horizontal movements (erosion and deposition), shown diagrammatically in Fig. 95. On this diagram the line ZOZ′ indicates coasts that are neither advancing nor retreating, either because emergence is cancelled out by erosion (ZO) or because submergence is cancelled out by deposition (OZ′): the point O represents an absolutely static coast on which no changes of any kind are in progress. Changes are most marked towards A, where emergence accompanied by deposition leads to rapid advance of the coast, and towards R, where erosion accompanying submergence results in rapid retreat. Rapid erosion may lead to recession of an emerging coast (X), while rapid deposition may prograde a coast that is being submerged (Y).

Bloom (1965) has suggested an elaboration of Valentin's scheme by the addition of a time-axis passing through O. On the three-dimensional diagram thus produced, it is possible to portray the

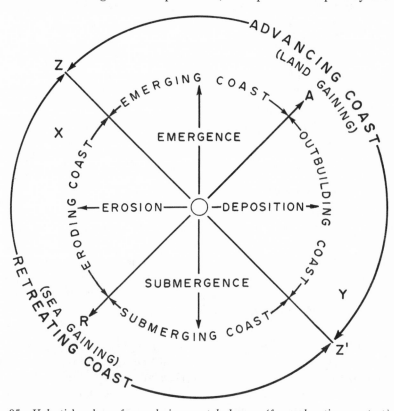

95 *Valentin's scheme for analysing coastal changes (for explanation, see text)*

course of evolution of a particular coast where relationships between emergence and submergence, erosion and deposition, have varied through time. An example is given from the Connecticut coast, where radiocarbon dates from buried peat horizons have yielded a chronology of relative changes of land and sea level in Recent times. At some stages the sea gained on the land during submergence, even though deposition continued; at other stages deposition was sufficiently rapid to prograde the land during continuing submergence. Now there is widespread erosion on the seaward margins of salt marshes, possibly because of resumed submergence. Each condition gives different points on Valentin's diagram, and Bloom's three-dimensional scheme offers a means of portraying coastal evolution as a wavy line drawn from a selected origin on the time axis and linking successive positions on the intersected planes.

Valentin attempted to measure present-day emergence and submergence from tide gauge records, accepting the evidence for a world-wide eustatic rise of sea level in progress at a rate of about 1 mm per year. Where mean tide levels are rising more rapidly than this, submergence is being accelerated by tectonic subsidence; where they are rising less rapidly, and where they are falling, submergence is being partly offset by tectonic uplift and the coast is emerging. Tide gauge records are not available for all sections of coast, and this kind of analysis remains somewhat speculative. Valentin's map showing present-day coastal evolution distinguished emerging, prograding, submerging, and retrograding coasts, and gave figures for changes of level where possible.

Valentin's scheme was a definite advance on earlier classifications made in terms of theoretical 'marine cycles', initiated by submergence or emergence as propounded by Johnson in terms of the Davisian 'cycle' concept (Davis, 1912), and reproduced in the older textbooks. This cyclic hypothesis has now outlived its usefulness, even as a teaching device, and can be seen in retrospect to have stultified coastal research. The essential idea was that a coast formed by submergence, initially indented, would pass through stages of 'youth' as promontories were cut back and sediment derived from them built laterally into spits and barriers enclosing embayments, 'sub-maturity' as the barriers retreated and coastal erosion drove back the headlands, and 'maturity' when all traces of the original indentations were destroyed, and the barriers came to rest as beaches along a smoothed, cliffed coast. It is quite possible that certain coasts have evolved in this way, but the hypothetical sequence cannot be accepted as 'normal' or 'typical'. The objections are that newly-submerged coasts are not necessarily indented; that crenulation by marine erosion can be a prolonged phase, not necessarily followed by the smoothing and simplification of coastal outlines; that where spits and barriers have formed the sediment is not necessarily derived from erosion of intervening cliffed headlands; that spits and barriers, formed on a submerged coast, are not necessarily driven landwards, and will prograde seawards if sediment supply is abundant; that the theoretical scheme is not a 'cycle', returning to the initial form, but a postulated 'sequence', the general validity of which has never been demonstrated; and that such idealised theoretical schemes of coastal evolution prompt facile interpretation of observed coastal features, which appear to fit the theoretical sequence, and are accepted as evidence of it without proper investigation of their geomorphological evolution. Similar objections apply to the hypothetical sequence of evolution of coasts following emergence, and neither scheme is satisfactory unless the coast becomes stable, following submergence or emergence. The advantage of 'non-cyclic' classifications, such as those proposed by Valentin and Bloom, is that they pose problems and stimulate further research instead of fitting observed features into presupposed evolutionary sequences. The liberation of coastal geomorphology from premature assumptions of 'cyclic' development was timely; it has been taken a stage farther by McGill (1958) in compiling a map of the coastal landforms of the world.

McGill's map (scale 1:25,000,000) shows the major landforms of the coastal fringe, 5–10 miles (8–16 km) wide, with selected shore features and certain other information relevant to analysis of coastal evolution. It is a useful aid to research, prompting further inquiry and comparative study of coastal landforms as a means of isolating and measuring the factors at work.

Major coastal landforms are classified in terms of their hinterlands as follows:

I.   Lowlands:

    A. Constructional (depositional) plains:

        1. dominantly flat-layered structure:

           a. existing ice

    b. plains of glacial deposition
    c. plains of fluvial deposition (including deltas)
    d. dune plains
    e. coral flats
    f. lava plains

B. Destructional (erosional) plains:
    1. flat-layered structure:
        a. sedimentary plains formed by glaciation
        b. sedimentary plains formed by fluvial erosion
        c. volcanic plains formed by glaciation
        d. volcanic plains formed by fluvial erosion
    2. complex structure:
        a. formed by glaciation
        b. formed by fluvial erosion

II. Uplands:

A. Plateaus:
    1. flat-layered structure:
        a. glaciated sedimentary plateau
        b. fluvially-eroded sedimentary plateau
        c. glaciated volcanic plateau
        d. fluvially-eroded volcanic plateau
    2. complex structure:
        a. formed by glaciation
        b. formed by fluvial erosion

B. Hills:
    1. flat-layered structure, remnant hills:
        a. glaciated sedimentary plateau
        b. fluvially-eroded sedimentary plateau
        c. glaciated volcanic plateau
        d. fluvially-eroded volcanic plateau
    2. complex structure:
        a. formed by glaciation
        b. formed by fluvial erosion

C. Mountains: subdivisions as for hills

D. Constructional (depositional) uplands:
        a. ice plateau
        b. dune hills

    c.　elevated coral flats

    d.　lava plateau

    e.　volcanoes (active and inactive).

This information is supplemented by indications of selected shore features, constructional or destructional, in the backshore, foreshore, and offshore zones, classified by the agent responsible: sea, wind, coral, or vegetation. This enables such features as backshore dunes, tidal flats, sedimentary and coral barriers, and 'tidal woodlands' (usually mangrove swamps) to be indicated. Finally, the map indicates such relevant features as:

    1.　the limits of present-day glaciation and the maximum extent of ice during the Last Glacial phase;

    2.　the approximate limit of permafrost;

    3.　the average limit of polar sea ice, with spring maximum and autumn minimum of pack ice;

    4.　the approximate limit of reef coral (here taken as the 20°C isotherm of sea water in the coldest month);

    5.　tide ranges larger than 10, and larger than 20 feet, both general and where localised in embayments;

    6.　certain characteristics of coastal embayments, their form and orientation to the coastal margin;

    7.　the coastal areas affected by isostatic recovery following deglaciation; and

    8.　the location of karst topography on limestone coasts.

The map was published by the American Geographical Society in the *Geographical Review* for 1958, and is worth close examination. It stands as a challenge to further investigation of coastal landforms in many parts of the world, and to the making of more detailed morphological maps of coastal features. The American Geographical Society has also published 1:50,000,000 maps of the world showing coastal climates (Bailey, 1958) and coastal vegetation (Axelrod, 1958), which give a valuable background to the analysis of coastal morphogenic systems, and these could also be developed on a local scale for research on particular coasts. There is a need for many more detailed studies of coastal landforms and their environments, as a basis for a more accurate and more comprehensive understanding of the world's coasts.

SELECTED BIBLIOGRAPHY

The scientific literature on coastal geomorphology is extensive, and is currently increasing at the rate of about four hundred publications a year. This selected bibliography of works mentioned in the text is intended as an introduction to the research literature. Advanced students will need to consult bibliographies issued every four years by the International Geographical Union Commission on Coastal Geomorphology, and bibliographies published by the Commission on Shorelines of I.N.Q.U.A. (International Association for Quaternary Research), as well as the relevant sections of *Geographical Abstracts A (Geomorphology)*, published by K. M. Clayton (University of East Anglia, Norwich, England). Useful material will also be found in *Coastal Research Notes*, published quarterly by the Geology Department of Florida State University, Tallahassee, U.S.A.

AGASSIZ, A. 1893. *The Great Barrier Reef of Australia*. London.

——1903. On the formation of barrier reefs and of the different types of atolls. *Proc. R. Soc.* **71**: 412–14.

AHNERT, F. 1960. Estuarine meanders in the Chesapeake Bay area. *Geogrl Rev.* **50**: 390–401.

ALLEN, J. R. L. 1965. Coastal geomorphology of eastern Nigeria: beach-ridge barrier islands and vegetated tidal flats. *Geologie Mijnb.* **44**: 1–21.

——and WELLS, J. W. 1962. Holocene coral banks and subsidence in the Niger delta. *J. Geol.* **70**: 381–97.

ARBER, M. A. 1940. The coastal landslips of south-east Devon. *Proc. Geol. Ass.* **51**: 257–71.

ARTHUR, R. S., MUNK, W. H., and ISAACS, J. D. 1952. The direct construction of wave rays. *Trans. Am. geophys. Un.* **33**: 855–65.

AXELROD, D. I. 1958. Coastal vegetation of the world. Map issued with *Second Coastal Geography Conference* (ed. R. J. Russell). Washington.

BAGNOLD, R. A. 1947. Sand movements by waves: some small-scale experiments with sand of very low density. *J. Instn civ. Engrs* **27**: 457–69.

BAILEY, H. P. 1958. An analysis of coastal climates, with particular reference to humid mid-latitudes. *Second Coastal Geography Conference* (ed. R. J. Russell). Washington: 23–56.

BAKER, G. 1956. Sand drift at Portland Harbour, Victoria. *Proc. R. Soc. Vict.* **68**: 151–98.

——1958. Stripped zones at cliff edges along a high wave energy coast, Port Campbell, Victoria. *Proc. R. Soc. Vict.* **71**: 175–9.

——and AHMAD, N. 1959. Re-examination of the fjord theory of Port Davey, Tasmania. *Proc. R. Soc. Tasmania* **93**: 113–15.

227

BASCOM, W. N. 1951. Relationship between sand size and beach face slope. *Trans. Am. geophys. Un.* **32**: 866–74.

——1954. The control of stream outlets by wave refraction. *J. Geol.* **62**: 600–5.

——1959. Ocean waves. *Scient. Am.* **201**: 74–84.

BAUER, F. H. 1961. Chronic problems of terrace study in southern Australia. *Z. Geomorph.* Supp. 3: 57–72.

BAULIG, H. 1956. *Vocabulaire Franco-Anglo-Allemand de Géomorphologie.* Strasbourg.

BEACH EROSION BOARD. 1961. *Shore Protection Planning and Design.* Washington.

BENSON, W. N. 1963. Tidal colks in Australia and New Zealand. *N.Z. Jl Geol. Geophys.* **6**: 634–40.

BIEDERMAN, E. W. 1962. Distinction of shoreline environments in New Jersey. *J. sedim. Petrol.* **32**: 181–200.

BIRD, E. C. F. 1965. *A Geomorphological Study of the Gippsland Lakes.* Australian National University Dept of Geography Publication G/1. Canberra.

——1967a. Coastal lagoons of southeastern Australia. *Landform Studies from Australia and New Guinea* (eds. J. N. Jennings, J. A. Mabbutt): 366–85.

——1967b. Depositional features in estuaries on the south coast of New South Wales. *Australian Geographical Studies* **5**: 113–24.

——and DENT, O. F. 1965. Shore platforms on the south coast of New South Wales. *Aust. Geogr.* **10**: 71–80.

——and RANWELL, D. S. 1964. The physiography of Poole Harbour, Dorset. *J. Ecol.* **52**: 355–66.

BLANC, J. J. 1954. Sédimentation dans la baie de Porto-Vecchio (Corse). *Revue. Géomorph. dyn.* **5**: 2–18.

BLOOM, A. L. 1965. The explanatory description of coasts. *Z. Geomorph.* **9**: 422–36.

BOURCART, J. 1952. *Les Frontières de l'Océan.* Paris.

BOWLER, J. 1966. Geology and geomorphology. Port Phillip Bay survey 1957–63. *Mem. natn. Mus. Vict.* **27**: 19–67.

BRADLEY, W. 1958. Submarine abrasion and wave-cut platforms. *Bull. geol. Soc. Am.* **69**: 967–74.

BRIQUET, A. 1923. Les dunes littorales. *Annls Géogr.* **32**: 385–94.

BROTHERS, R. N. 1954. A physiographical study of Recent sand dunes on the Auckland west coast. *N.Z. Geogr.* **10**: 47–59.

BRUUN, P. 1962. Sea level rise as a cause of shore erosion. *Proc. Am. Soc. civ. Engrs* (Waterways and Harbours Divn) **88**: 117–30.

——1967. *Tidal Inlets and Littoral Drift.* Oslo.

——and GERRITSEN, F. 1959. Natural by-passing of sand at coastal inlets. *Proc. Am. Soc. civ. Engrs* (Waterways and Harbours Divn) **85**: 75–107.

BURGES, A., and DROVER, D. P. 1953. The rate of podzol development in the sands of the Woy Woy district, N.S.W. *Aust. J. Bot.* **1**: 83–94.

CARR, A. P. 1962. Cartographic record and historical accuracy. *Geography* **47**: 135–44.

——1965. Shingle spit and river mouth: short-term dynamics. *Trans. Inst. Br. Geogr.* **36**: 117–30.

CASTANY, G., and OTTMANN, F. 1957. Le Quaternaire marin de la Méditerranée occidentale. *Revue Géogr. phys. Géol. dyn.* **1**: 46–55.

CHAPMAN, V. J. 1960. *Salt Marshes and Salt Deserts of the World.* London.

CHAPPELL, J. 1967. Recognizing fossil strand lines from grain size analysis. *J. sedim. Petrol.* **37**: 157–65.

CHATERJEE, S. P. 1961. Fluctuations of sea level around the coasts of India during the Quaternary period. *Z. Geomorph.* Supp. **3**: 48–56.

CHURCHILL, D. M. 1959. Late Quaternary eustatic changes in the Swan River district. *J. Proc. R. Soc. West Aust.* **42**: 53–5.

COALDRAKE, J. E. 1962. The coastal sand dunes of southern Queensland. *Proc. R. Soc. Qd* **72**: 101–16.

COLEMAN, J. M., and SMITH, W. G. 1964. Late Recent rise of sea level. *Bull. geol. Soc. Am.* **75**: 833–40.

COOPER, W. S. 1958. Coastal sand dunes of Oregon and Washington. *Mem. geol. Soc. Am.* **72**.

COTTON, C. A. 1942. Shorelines of transverse deformation. *J. Geomorph.* **5**: 45–58.

———1951a. Une côte de déformation transverse à Wellington (N.Z.). *Revue Géomorph. dyn.* **2**: 97–109.

———1951b. Sea cliffs of Banks Peninsula and Wellington: some criteria for coastal classification. *N.Z. Geogr.* **7**: 103–20.

———1952. Criteria for the classification of coasts. *Proc. 17th Conference of the International Geographical Union* (Washington): 315–19.

———1954. Tests of a German non-cyclic theory and classification of coasts. *Geogrl J.* **120**: 353–61.

———1956. Rias *sensu stricto* and *sensu lato. Geogrl J.* **122**: 360–4.

———1963. Levels of planation of marine benches. *Z. Geomorph.* **7**: 97–111.

CURRAY, J. R. 1960. Sediments and history of Holocene transgression, continental shelf, northwest Gulf of Mexico. *Recent Sediments, northwest Gulf of Mexico* (Tulsa, Okla.): 221–66.

DALY, R. A. 1910. Pleistocene glaciation and the coral reef problem. *Amer. J. Sci.* **30**: 297–308.

———1934. *The Changing World of the Ice Age.* Yale.

DARBYSHIRE, J. 1961. Prediction of wave characteristics over the North Atlantic. *J. Inst. Navig.* **14**: 339–47.

DARWIN, C. 1842. *The Structure and Distribution of Coral Reefs.* London.

DAVID, T. W. E. 1950. *The Geology of the Commonwealth of Australia* (ed. W. R. Browne). London.

DAVIES, J. L. 1957. The importance of cut and fill in the development of sand beach ridges. *Aust. J. Sci.* **20**: 105–11.

———1959. Wave refraction and the evolution of shoreline curves. *Geogrl Stud.* **5**: 1–14.

———1964. A morphogenic approach to world shorelines. *Z. Geomorph.* **8** (Mortensen Sonderheft): 127–42.

DAVIS, W. M. 1896. The outline of Cape Cod. *Proc. Am. Acad. Arts Sci.* **31**: 303–32.

———1928. *The Coral Reef Problem.* American Geographical Society Special Publication 9.

——and Rühl, A. 1912. *Die erklärende Beschreibung der Landformen.* Leipzig.

Day, J. H. 1951. The ecology of South African estuaries, I: a review of estuarine conditions in general. *Trans. R. Soc. S. Afr.* **33**: 53–91.

Derrington, S. S. 1960. *Completion Report, No. 1 Bore Wreck Island.* Bureau of Mineral Resources Petroleum Search Publication 4.

Dietz, R. S. 1963. Wave-base, marine profile of equilibrium, and wave-built terraces: a critical appraisal. *Bull. geol. Soc. Am.* **74**: 971–90.

Dobrin, M. B., Perkins, B., and Snaveley, B. L. 1949. Sub-surface constitution of Bikini Atoll as indicated by a seismic-refraction survey. *Bull. geol. Soc. Am.* **60**: 807–28.

Donn, W. L., Farrand, W. R., and Ewing, M. 1962. Pleistocene ice volumes and sea level lowering. *J. Geol.* **70**: 206–14.

——and Shaw, D. M. 1963. Sea level and climates of the past century. *Science* **142**: 1166–7.

Donovan, D. T., and Stride, A. H. 1961. Erosion of a rock floor by tidal streams. *Geol. Mag.* **98**: 393–8.

Dunbar, G. S. 1956. *Geographical History of the Carolina Banks.* Coastal Studies Institute, Louisiana State University, Technical Report 8.

Edwards, A. B. 1941. Storm-wave platforms. *J. Geomorph.* **4**: 223–36.

——1958. Wave-cut platforms at Yampi Sound in the Buccaneer Archipelago. *J. Proc. R. Soc. West Aust.* **41**: 17–21.

Emery, K. O. 1960. *The Sea off Southern California.* New York.

——1961. Submerged marine terraces and their sediments. *Z. Geomorph.* Supp. **3**: 17–29.

——and Stevenson, R. E. 1957. Estuaries and lagoons. *Mem. geol. Surv. America* **67**: 673–750.

Emiliani, C. 1963. Deep-sea sediments. *Trans. Am. geophys. Un.* **44**: 495–8.

Evans, O. F. 1942. The origin of spits, bars, and related structures. *J. Geol.* **50**: 846–63.

Everard, C. E. 1960. Mining and shoreline evolution near St Austell, Cornwall. *Trans. R. geol. Soc. Corn.* **19**: 199–219.

Fairbridge, R. W. 1947. A contemporary eustatic rise in sea level? *Geogrl J.* **109**: 157.

——1948. Notes on the geomorphology of the Pelsart Group of the Houtman's Abrolhos Islands. *J. Proc. R. Soc. West Aust.* **33**: 1–43.

——1950. The geology and geomorphology of Point Peron, W. A. *J. Proc. R. Soc. West Aust.* **34**: 35–72.

——1952. The Sahul Shelf, Northern Australia: its structure and geological relationships. *J. Proc. R. Soc. West Aust.* **37**: 1–34.

——1961. Eustatic changes in sea level. *Physics and Chemistry of the Earth* **4**: 99–185.

——1967. Coral reefs of the Australian region. *Landform Studies from Australia and New Guinea* (eds. J. N. Jennings, J. A. Mabbutt): 386–417.

——and Stewart, H. B. 1960. Alexa Bank, a drowned atoll on the Melanesian border plateau. *Deep Sea Res.* **7**: 100–16.

——and Teichert, C. 1948. The Low Isles of the Great Barrier Reef: a new analysis. *Geogrl J.* **111**: 67–88.

——and—— 1952. Soil horizons and marine bands in the coastal limestones of Western Australia. *J. Proc. R. Soc. West Aust.* **86**: 68–87.

FISK, H. N. 1958. Padre Island and the Laguna Madre flats, coastal south Texas. *Second Coastal Geography Conference:* 103–51.

——and McFARLAN, E. 1955. *Late Quaternary Deltaic Deposits of the Mississippi River.* Geological Society of America Special Publication **24**: 279–302.

FLEMING, C. A. 1965. Two-storied cliffs at the Auckland Islands. *Trans. R. Soc. N.Z. (Geol.)* **3**: 171–4.

FLINT, R. F. 1966. Comparison of interglacial marine stratigraphy in Virginia, Alaska, and Mediterranean areas. *Am. J. Sci.* **264**: 673–84.

FOLK, R. L., and ROBLES, R. 1964. Carbonate sands of Isla Perez, Alacran Reef complex, Yucatan. *J. Geol.* **72**: 255–92.

FRIEDMAN, G. M. 1961. Distinction between dune, beach, and river sands from their textural characteristics. *J. sedim. Petrol.* **31**: 514–29.

FUENZALIDA, H. *et al.* 1965. *High Stands of Quaternary Sea Level along the Chilean Coast.* Geological Society of America Special Publication **84**: 473–96.

GARDNER, D. 1955. Beach sand heavy mineral deposits of eastern Australia. *Bureau of Mineral Resources Bulletin* (Canberra) **28**.

GIERLOFF-EMDEN, H. G. 1959. Lagunen, Nehrungen, Strandwälle und Flussmündungen im gesehen tropische Flachlandküsten. *Z. Geomorph.* **3**: 29–46.

——1961. Nehrungen und Lagunen. *Petermanns geogr. Mitt.* **105**: 81–92; 161–76.

GILBERT, G. K. 1890. *Lake Bonneville.* U.S. Geological Survey Monograph 1.

GILL, E. D. 1961. Changes in the level of the sea relative to the level of the land in Australia. *Z. Geomorph.* Supp. **3**: 73–9.

GRESSWELL, R. K. 1953. *Sandy Shores in South Lancashire.* Liverpool.

GUILCHER, A. 1953. Essai sur la zonation et la distribution des formes littorales de dissolution du calcaire. *Annls Géogr.* **62**: 161–79.

——1956. Les travaux de Berthois sur l'estuaire de la Loire. *Norois* **3**: 87–90 and **4**: 106–7.

——1958a. *Coastal and Submarine Morphology* (eds. B. W. Sparks, R. H. W. Kneese). London.

——1958b. Coastal corrosion forms in limestone around the Bay of Biscay. *Scott. geogr. Mag.* **74**: 137–49.

——1963. Quelques caractères des récifs-barrières et de leurs lagons. *Bull. Ass. Géogr. fr.* **314–15**: 2–15.

——and BERTHOIS, L. 1957. Cinq années d'observations sédimentologiques dans quatre estuaires-témoins de l'ouest de la Bretagne. *Revue Géomorph. dyn.* **8**: 67–86.

——and NICOLAS, J. P. 1954. Observations sur la langue de Barbarie et les bras du Sénégal aux environs de St Louis. *Bull. Inf. Com. cent. Océanogr. Étude Côtes* **6**: 227–42.

GULLIVER, F. P. 1889. Shoreline topography. *Proc. Am. Acad. Arts Sci.* **34**: 151–258.

GUTENBERG, B. 1941. Changes in sea level, postglacial uplift and mobility of the earth's interior. *Bull. geol. Soc. Am.* **52**: 721–72.

HAILS, J. R. 1965. A critical review of sea-level changes in eastern Australia since the Last Glacial. *Australian Geographical Studies* **3**: 63–78.

HARRIS, R. L. 1955. *Restudy of Test Shore Nourishment by Offshore Deposition of Sand, Long Branch, New Jersey.* Beach Erosion Board Technical Memorandum 62.

HAUGHTON, S. H. 1963. *The Stratigraphic History of Africa South of the Sahara.* Edinburgh.

HELLE, J. R. 1958. *Surf Statistics for the Coasts of the United States.* Beach Erosion Board Technical Memorandum 108.

HILLS, E. S. 1949. Shore platforms. *Geol. Mag.* **86**: 137–52.

HJULSTRÖM, F. 1939. Transportation of detritus by moving water. *Recent Marine Sediments* (ed. P. D. Trask): 5–31.

HODGKIN, E. P. 1964. Rate of erosion of intertidal limestone. *Z. Geomorph.* **8**: 385–92.

HOSSFELD, P. S. 1965. Radiocarbon dating and palaeoecology of the Aitape fossil human remains. *Proc. R. Soc. Vict.* **78**: 161–5.

HUBBARD, J. C. E. 1965. Spartina salt marshes in southern England, VI: Pattern of invasion in Poole Harbour. *J. Ecol.* **53**: 799–813.

JENNINGS, J. N. 1955. The influence of wave action on coastal outline in plan. *Aust. Geogr.* **6**: 36–44.

——1957a. On the orientation of parabolic or U-dunes. *Geogrl J.* **123**: 474–80.

——1957b. Coastal dune lakes as exemplified from King Island, Tasmania. *Geogrl J.* **123**: 59–70.

——1959. The submarine topography of Bass Strait. *Proc. R. Soc. Vict.* **71**: 49–72.

——1961. Sea level changes in King Island, Bass Strait. *Z. Geomorph.* Supp. **3**: 80–4.

——1964. The question of coastal dunes in tropical humid climates. *Z. Geomorph.* **8** (Mortensen Sonderheft): 150–4.

——1965. Further discussion of factors affecting coastal dune formation in the tropics. *Aust. J. Sci.* **28**: 166–7.

——1967. Cliff-top dunes. *Australian Geographical Studies* **5**: 40–9.

——and BIRD, E. C. F. 1967. Regional geomorphological characteristics of some Australian estuaries. *Estuaries* (ed. G. H. Lauff): 121–8.

JOHNSON, D. W. 1919. *Shore Processes and Shoreline Development.* New York.

——1925. *The New England-Acadian Shoreline.* New York.

——1931. Supposed two-metre eustatic bench of the Pacific shores. *Comptes Rendus, International Geographical Congress* (Paris) **2**: 158–63.

JOHNSON, J. W. 1956. Dynamics of nearshore sediment movement. *Bull. Am. Ass. Petrol. Geol.* **40**: 2211–32.

——1959. The littoral drift problem at shoreline harbours. *Proc. Am. Soc. civ. Engrs* (Waterways and Harbours Divn) **124**: 525–55.

——, O'BRIEN, M. P., and ISAACS, J. D. 1948. *Graphical Construction of Wave Refraction Diagrams.* U.S. Navy Hydrographic Office Publication 605.

JOLLIFFE, I. P. 1961. The use of tracers to study beach movements and the measurement of littoral drift by a fluorescent technique. *Revue Géomorph. dyn.* **12**: 81–95.

JUTSON, J. T. 1939. Shore platforms near Sydney, N.S.W. *J. Geomorph.* **2**: 237–50.

KAYE, C. A. 1959. Shoreline features and Quaternary shoreline changes, Puerto Rico. *U.S. Geological Survey Professional Paper* 317–B: 49–140.

KIDSON, C. 1963. The growth of sand and shingle spits across estuaries. *Z. Geomorph.* **7**: 1–22.

——and CARR, A. P. 1962. Marking beach materials for tracing experiments. *Proc. Am. Soc. civ. Engrs* (Waterways and Harbours Divn) **88**: 43–60.

KING, C. A. M. 1953. The relationship between wave incidence, wind direction and beach changes at Marsden Bay, County Durham. *Trans. Inst. Br. Geogr.* **19**: 13–23.

——1959. *Beaches and Coasts.* London.

——and BARNES, F. A. 1964. Changes in the configuration of the intertidal beach zone on part of the Lincolnshire coast since 1951. *Z. Geomorph.* **8**: 105–26.

——and WILLIAMS, W. W. 1949. The formation and movement of sand bars by wave action. *Geogrl J.* **113**: 70–85.

KING, L. C. 1962. *The Morphology of the Earth.* Edinburgh.

KRUMBEIN, W. C. 1963. *The Analysis of Observational Data from Natural Beaches.* Beach Erosion Board Technical Memorandum 130.

LANDON, R. E. 1930. An analysis of beach pebble abrasion and transportation. *J. Geol.* **38**: 437–46.

LAUFF, G. H. (ed.) 1967. *Estuaries.* Publication of the American Association for the Advancement of Science 83.

LAW, P. 1967. Geography in the Antarctic. *Aust. Geogr.* **10**: 145–51.

LE BOURDIEC, P. 1958. Aspects de la morphogénèse plio-quaternaire en basse Côte d'Ivoire. *Revue Géomorph. dyn.* **9**: 33–42.

LEWIS, W. V. 1932. The formation of Dungeness foreland. *Geogrl J.* **80**: 309–24.

——1938. The evolution of shoreline curves. *Proc. Geol. Ass.* **49**: 107–27.

——and BALCHIN, W. G. V. 1940. Past sea levels at Dungeness. *Geogrl J.* **96**: 258–85.

LONGUET-HIGGINS, M. S., and PARKIN, D. W. 1962. Sea waves and beach cusps *Geogrl J.* **128**: 195–201.

LUCKE, J. B. 1934. A theory of evolution of lagoon deposits on shorelines of emergence. *J. Geol.* **42**: 561–84.

McGILL, J. T. 1958. Map of coastal landforms of the world. *Geogrl Rev.* **48**: 402–5.

McKENZIE, P. 1958. Rip current systems. *J. Geol.* **66**: 103–13.

MARSHALL, P. 1929. Beach gravels and sands. *Trans. N.Z. Inst.* **60**: 324–65.

MASON, C. C., and FOLK, R. L. 1958. Differentiation of beach, dune, and aeolian flat environments by size analysis, Mustang Island, Texas. *J. Sedim. Petrol.* **28**: 211–26.

MILLER, D. J. 1960. Giant waves in Lituya Bay, Alaska. U.S. Geological Survey Professional Paper 354–C: 51–86.

MILLER, R. L. and ZEIGLER, J. M. 1958. A model relating dynamics and sediment pattern in equilibrium in the region of shoaling waves, breaker zone, and foreshore. *J. Geol.* **66**: 417–41.

MØLLER, J. T. 1963. Accumulation and abrasion in a tidal area. *Geogr. Tidsskr.* **62**: 56–79.

MOSS, A. J. 1962. The physical nature of common sandy and pebbly deposits. *Am. J. Sci.* **260**: 337–73 and **261**: 297–343.

MUNK, W. H. *et al.* 1963. Directional recording of waves from distant storms. *Phil. Trans. R. Soc.* Series A. **255**: 505–84.

——and TRAYLOR, M. A. 1947. Refraction of ocean waves: a process linking underwater topography to beach erosion. *J. Geol.* **55**: 1–26.

MURRAY, J. 1880. On the structure and origin of coral reefs and islands. *Proc. R. Soc. Edinb.* **10**: 505–18.

NICHOLS, R. L. 1948. Flying bars. *Amer. J. Sci.* **246**: 96–100.

——1961. Characteristics of beaches formed in polar climates. *Amer. J. Sci.* **259**: 694–708.

NOSSIN, J. J. 1965a. The geomorphic history of the northern Padang delta. *J. trop. Geogr.* **20**: 54–64.

——1965b. Analysis of younger beach ridge deposits in eastern Malaya. *Z. Geomorph.* **9**: 186–208.

OAKS, R. A. and COCH, N. K. 1963. Pleistocene sea levels, southeastern Virginia. *Science* **140**: 979–84.

OLSON, J. S. 1958. Lake Michigan dune development. *J. Geol.* **66**: 254–63, 345–51, and 473–83.

PICARD, J. 1954. Les 'schorres' de l'estuaire du Stabiacco (Porto-Vecchio). *Revue Géomorph. dyn.* **5**: 19–24.

PRÊCHEUR, C. 1960. *Le Littoral de la Manche: de Ste Adresse à Ault. Étude morphologique. Norois* special volume.

PRICE, W. A. 1947. Equilibrium of form and forces in tidal basins of the coast of Texas and Louisiana. *Bull. Am. Ass. Petrol. Geol.* **31**: 1619–63.

——1955. Environment and formation of chenier plain. *Quaternaria* **2**: 75–86.

PSUTY, N. P. 1965. Beach-ridge development in Tabasco, Mexico. *Ann. Ass. Am. Geogr.* **55**: 112–24.

PUTNAM, W. C. 1937. Marine cycle of erosion for steeply-sloping shoreline of emergence. *J. Geol.* **45**: 844–50.

REVELLE, R. and EMERY, K. O. 1957. Chemical erosion of beach rock and exposed reef rock. U.S. Geological Survey Professional Paper 260–T: 699–709.

RICHARDS, H. G. and FAIRBRIDGE, R. W. 1965. *Annotated Bibliography of Quaternary Shorelines, 1945–64.* Publication No. 6 of the Academy of Natural Science, Philadelphia.

ROBINSON, A. H. W. 1955. The harbour entrances of Poole, Christchurch and Pagham. *Geogrl J.* **121**: 33–50.

——1960. Ebb-flood channel systems in sandy bays and estuaries. *Geography* **45**: 183–99.

——1961. The hydrography of Start Bay and its relationship to beach changes at Hallsands. *Geogrl J.* **121**: 63–77.

——1966. Residual currents in relation to shoreline evolution of the East Anglian coast. *Marine Geology* **4**: 57–84.

RUSSELL, R. C. H., and MACMILLAN, D. H. 1954. *Waves and Tides.* London.

RUSSELL, R. J. 1958. Long straight beaches. *Eclog. geol. Helv.* **51**: 591–8.

——(ed.) 1961. Pacific island terraces: eustatic? *Z. Geomorph.* Supp. **3**.

——1962. Origin of beach rock. *Z. Geomorph.* **6**: 1–16.

——1963. Recent recession of tropical cliffy coasts. *Science* **139**: 9–15.

——1964. Techniques of eustacy studies. *Z. Geomorph.* **8** (Mortensen Sonderheft): 25–42.

——1966. Coral cap of Barbados. *Tidsch. K. ned. aardrijksk. Genoot.* **83**: 298–302.

——and HOWE, H. V. W. 1935. Cheniers of southwestern Louisiana. *Geogrl Rev.* **25**: 449–61.

——and MCINTYRE, W. G. 1965a. Beach cusps. *Bull. geol. Soc. Am.* **76**: 307–20.

——and——1965b. Southern hemisphere beach rock. *Geogrl Rev.* **55**: 17–45.

——and——1966. *Australian Tidal Flats.* Louisiana State University Coastal Studies Series 13.

——and RUSSELL, R. D. 1939. Mississippi River delta sedimentation. *Recent Marine Sediments* (ed. P. D. Trask): 153–77.

SAMOJLOV, I. V. 1956. *Die Flussmündungen.* Gotha.

SAVIGEAR, R. A. G. 1952. Some observations on slope development in South Wales. *Trans. Inst. Br. Geogr.* **18**: 31–51.

SAVILLE, T. 1950. Model study of sand transport along an infinitely long straight beach. *Trans. Am. geophys. Un.* **31**: 555–65.

SCHOU, A. 1945. *Det Marine Forland.* Copenhagen.

——1952. Direction determining influence of the wind on shoreline simplification and coastal dunes. *Proc. 17th Conference of the International Geographical Union.* Washington: 370–3.

SCHWARTZ, M. L. 1967. The Bruun theory of sea level rise as a cause of shore erosion. *J. Geol.* **75**: 76–92.

SHEPARD, F. P. 1937. Revised classification of marine shorelines. *J. Geol.* **45**: 602–24.

——1948. *Submarine Geology.* New York.

——1960. Gulf Coast barriers. *Recent Sediments, northwest Gulf of Mexico* (Tulsa, Okla.): 197–220.

——1963. *Submarine Geology.* New York (2nd ed.).

——, EMERY, K. O., and LA FOND, E. C. 1941. Rip currents: a process of geological importance. *J. Geol.* **49**: 337–69.

——and INMAN, D. L. 1950. Nearshore water circulation related to bottom topography and wave refraction. *Trans. Am. geophys. Un.* **31**: 196–212.

——and YOUNG, R. 1961. Distinguishing between beach and dune sands. *J. sedim Petrol.* **31**: 196–214.

SILVESTER, R. 1956. The use of cyclonicity charts in the study of littoral drift. *Trans. Am. geophys. Un.* **37**: 694–6.

SMITH, T. H. and IREDALE, T. 1924. Evidence of a negative movement of the strand line of 400 feet in New South Wales. *J. Proc. R. Soc. N.S.W.* **58**: 157–68.

SPENDER, M. 1930. Island-reefs of the Queensland coast. *Geogrl J.* **76**: 193–214 and 273–93.

SPRIGG, R. C. 1959. Stranded sea beaches and associated sand accumulations of the upper south-east of South Australia. *Trans. R. Soc. S. Aust.* **82**: 183–93.

STEARNS, H. T. 1961. Eustatic shorelines on Pacific islands. *Z. Geomorph.* Supp. **3**: 3–16.

STEERS, J. A. 1929. The Queensland coast and the Great Barrier Reefs. *Geogrl J.* **74**: 232–57 and 341–70.

——1937. The coral islands and associated features of the Great Barrier Reefs. *Geogrl J.* **89**: 1–28. and 118–46.

——1953a. *The Sea Coast*. London.

——1953b. The east coast floods, January 31–February 1, 1953. *Geogrl J.* **119**: 280–98.

——(ed.) 1960. *Scolt Head Island*. Cambridge.

——1964. *The Coastline of England and Wales*. Cambridge.

STEPHENS, N. and SYNGE, F. M. 1966. Pleistocene shorelines. *Essays in Geomorphology* (ed. G. H. Dury): 1–51.

STEVENSON, R. E. and EMERY, K. O. 1958. *Marshlands at Newport Bay, California*. Allan Hancock Foundation Publication 20.

STIPP, J. J., CHAPPELL, J. M. A., and McDOUGALL, I. 1967. K/Ar age estimate of the Pliocene-Pleistocene boundary in New Zealand. *Am. J. Sci.* **265**: 462–74.

STODDART, D. R. 1965a. British Honduras cays and the low wooded island problem. *Trans. Inst. Br. Geogr.* **36**: 131–48.

——1965b. Re-survey of hurricane effects on the British Honduras reefs and cays. *Nature, Lond.* **207**: 589–92.

——and CANN, J. R. 1965. Nature and origin of beach rock. *J. sedim. Petrol.* **35**: 243–7.

SUESS, E. 1906. *The Face of the Earth* (trans. W. J. Sollas). Oxford.

TANNER, W. F. 1963. Origin and maintenance of ripple marks. *Sedimentology* **2**: 307–11.

——, EVANS, R. G., and HOLMES, C. W. 1963. Low-energy coast near Cape Romano, Florida. *J. sedim. Petrol.* **33**: 713–22.

TEICHERT, C. 1950. Late Quaternary changes of sea level at Rottnest Island, W. A. *Proc. R. Soc. Vict.* **59**: 63–78.

THOM, B. G. 1964. Origin of sand beach ridges. *Aust. J. Sci.* **26**: 351–2.

——1965. Late Quaternary sand deposits between Newcastle and Seal Rocks, New South Wales. *Proc. R. Soc. N.S.W.* **98**: 22–36.

TIETZE, W. 1962. A contribution to the geomorphological problem of strand-flats. *Petermanns geogr. Mitt.* **106**: 1–20.

TRASK, P. D. 1955. *Movement of Sand around Southern Californian Promontories*. Beach Erosion Board Technical Memorandum 76.

TRICART, J. 1956. Aspects morphologiques du delta du Sénégal. *Revue Géomorph. dyn.* **7**: 65–85.

——1959. Problèmes géomorphologiques du littoral oriental du Bresil. *Cah. océanogr.* **11**: 276–308.

——1962. Observations de géomorphologie littorale à Mamba Point, Liberia. *Erdkunde* **16**: 49–57.

TRICKER, R. A. R. 1964. *Bores, Breakers, Waves and Wakes*. London.

TROLL, C. and SCHMIDT-KRAEPELIN, E. 1965. Das neue Delta des Rio Sinu an der Karibischen Küste Kolumbiens. *Erdkunde* **19**: 14–23.

VALENTIN, H. 1952. Die Küsten der Erde. *Petermanns geogr. Mitt. (Erg.)* 246.

——1961. The central west coast of Cape York Peninsula. *Aust. Geogr.* **8**: 65–72.

VAN DIEREN, J. W. 1934. *Organogene Dünenbildung*. The Hague.

VAN STRAATEN, L. M. J. U. 1953. Megaripples in the Dutch Wadden Sea and in the Basin of Arcachon. *Geologie Mijnb.* **15**: 1–11.

——1959. Littoral and submarine morphology of the Rhône delta. *Second Coastal Geography Conference Proceedings:* 233–64.

——1965. Coastal barrier deposits in south and north Holland. *Meded. geol. Sticht.* **17**: 41–75.

VARNES, D. J. 1950. Relation of landslides to sedimentary features. *Applied Sedimentology* (ed. P. D. Trask): 229–46.

VERSTAPPEN, H. 1953. *Djakarta Bay: A Geomorphological Study in Shoreline Development.* La Haye, Ohio.

VLADIMIROV, A. T. 1961. The morphology and evolution of the lagoon coast of Sakhalin. *Trudy Inst. Okeanol.* (Moscow). **48**: 145–71.

WALTON, K. (ed.) 1966. The Vertical Displacement of Shorelines in Highland Britain. *Trans. Inst. Br. Geogr.* **39**.

WEBB, J. E. 1958. The ecology of Lagos lagoon. *Phil. Trans. R. Soc.* **241**: 307–18.

WENTWORTH, J. K. 1939. Marine bench-forming processes: solution benching. *J. Geomorph.* **2**: 3–25.

WEXLER, H. 1961. Ice budgets for Antarctica and changes of sea level. *J. Glaciol.* **3**: 867–72.

WIENS, H. J. 1962. *Atoll Environment and Ecology.* Yale.

WILLIAMS, W. W. 1956. An east coast survey: some recent changes in the coast of East Anglia. *Geogrl J.* **122**: 317–34.

YASSO, W. E. 1965a. Plan geometry of headland-bay beaches. *J. Geol.* **73**: 702–14.

——1965b. Fluorescent tracer particle determination of the size-velocity relation for foreshore sediment transport, Sandy Hook, New Jersey. *J. sedim. Petrol.* **35**: 989–93.

——1966. Formulation and use of fluorescent tracer coatings in sediment transport studies. *Sedimentology* **6**: 287–301.

ZENKOVITCH, V. P. 1959. On the genesis of cuspate spits along lagoon shores. *J. Geol.* **67**: 269–77.

ZENKOVICH, V. P. 1967. *Processes of Coastal Development* (trans. O. G. Fry; ed. J. A. Steers). Edinburgh.

ZEUNER, F. E. 1959. *The Pleistocene Period.* London.

INDEX

Abrasion, 49, 63–7, 69, 72, 195; notch, 63; platform, 50; ramp, 63
Accretion: in salt marshes, 159, 160; on dunes, 129–31, 133
Aeolianite, *see* Calcarenite, aeolian
Agassiz, A., 194, 200
Aggradation, 187–8
Ahnert, F., 153
Aitape, 30
Alexa Bank, 199
Allen, J. R. L., 158
Allen, J. R. L. and Wells, J. W., 186
Ammophila arenaria, 129; *A. brevigulata*, 131
Antarctic ice sheet, 39, 47, 218–19
Apollo Bay, 100, 101
Apsley Strait, 22
Aral Sea, 169
Arber, M. A., 53
Arthur, R. S., Munk, W. H., and Isaacs, J. D., 18
Atolls, 190, 192, 197–8; Bikini, 74, 201; Christmas I., 206; Elizabeth, 190, 197; Kapingamarangi, 197; Mare, 206; Middleton, 190, 197; Scott Reefs, 191, 197; Seringapatam, 191, 197; Uvea, 206; compound, 197–8; drowned, 199; emerged, 206; oceanic, 197–8; shelf, 197
Attrition, 49, 88
Auckland Island, 57
Axelrod, D. I., 226
Axmouth-Dowlands landslip, 53, 218

Backshore, 1, 2
Backplain depression, 187
Backswamp depression, 187, 189
Backwash, 12, 83, 100, 195
Bagnold, R. A., 84
Bailey, H. P., 3, 226
Baker, G., 56, 80, 93–4
Baker, G. and Ahmad, N., 150
Banks Peninsula, 77, 123
Bantry Bay, 147
Bar, 2, 102, 165
Barnegat Inlet, 156
Barrenjoey, 114
Barrier reefs, 190, 192, 194; submerged, 203; *see also* Great Barrier Reefs
Barriers, coastal, 1, 2, 118–27, 161, 162, 169, 174–6, 182; barrier beach, 118; barrier island, 118; bay barrier, 118
Bascom, W. N., 10, 18, 105, 165
Bassin d'Arcachon, 141, 143, 164

Bass Strait, 43, 87
Bathurst Channel, 150, 151
Bauer, F. H., 136, 138
Bay of Biscay, 77
Bay of Fundy, 9
Bay of St Michel, 9
Beach, 1, 2, 81–107, 182, 184; Arctic and Antarctic coasts, 83, 107; compartments, 84, 89, 90; cusps, 106–7; drifting, 13, 88–92; dynamics, 85–6; erosion, 84, 90, 92, 101; gradients, 82, 83; lateral variations, 87, 88,105, 106; nourishment, 85, 101; outlines in plan, 18, 19, 95–100; plan geometry, 100; process-response model, 86; profile, 83, 100–4; response to sea level change, 101, 102; ridges, 103, 104, 123, 127; rock, 107, 108, 207, 209, 211; theoretical treatment, 85, 86; upper and lower, 99, 105
Beach sediments: composition of, 86–8; glacial erratics in, 40, 83, 107; mechanical analysis of, 81–2; origin of, 83–5; skewness, 81, 82; sorting, 81, 82, 86, 105–7
Beachy Head, 58
Benson, W. N., 23
Berm, 1, 100, 102
Biederman, E. W., 88
Biotic factor in coastal evolution, 3, 4, 64, 65, 75, 85, 107, 108, 128, 129, 138, 141, 142, 154–62, 166–70, 189, 190–3, 211
Bird, E. C. F., 57, 85, 86, 123, 127, 140, 161, 162, 165, 169, 189
Bird, E. C. F. and Dent, O. F., 67, 72
Bird, E. C. F. and Ranwell, D. S., 156
Blakeney Point, 20, 89, 99, 108, 109, 111
Blanc, J. J., 160
Blind estuary, 161
Bloom, A. L., 222, 224
Blowhole, 18, 19, 50
Blowout, 138–40
Bonifacio, 53
Boomer Beach, 89
Boston Harbour, 112, 217
Botany Bay, 107
Boulders, 81
Bourcart, J., 29
Bournemouth Bay, 55, 58, 61, 84
Bowler, J., 89
Bradley, W., 64
Braunton Burrows, 128

239